# Green Energy and Technology

For further volumes:
http://www.springer.com/series/8059

J. S. Shrimpton · S. Haeri · Stephen J. Scott

# Statistical Treatment of Turbulent Polydisperse Particle Systems

## A Non-sectional PDF Approach

Springer

J. S. Shrimpton
Faculty of Engineering and
the Environment
University of Southampton
Southampton
UK

Stephen J. Scott
York
UK

S. Haeri
University of Southampton
Southampton
UK

ISSN 1865-3529
ISBN 978-1-4471-6991-8
DOI 10.1007/978-1-4471-6344-2
Springer London Heidelberg New York Dordrecht

ISSN 1865-3537 (electronic)
ISBN 978-1-4471-6344-2 (eBook)

Printed on acid-free paper

Springer is part of Springer Science+Business Media (www.springer.com)

*To my wife Parvaneh for her patience and support*

*– Sina –*

# Preface

This monograph is an advanced text in fluid dynamics intended for researchers and graduate students in the field of multiphase flows or powder technology. In this monograph, a novel formulation strategy, based on the probability density function, for general multiphase flow systems is introduced and thoroughly discussed. The novelty of the current approach, compared to the other non-PDF-based methods, lies in the fact that formulation starts by writing the fundamental ordinary differential equations (ODEs) governing the motion of Lagrangian fluid and discrete particles. This allows physical models to be applied at that level.

The other objective is to apply the aforementioned strategy to the formulation of turbulent particle-flow systems without assuming a constant particle size. Although there have been several attempts to include the particle size distribution into the governing equations none of these previous formulations can coherently incorporate the size distribution into the governing equations. However, it is shown in this monograph how particle size can be added naturally to the equations by first writing the fundamental ODEs governing the motion of the individual fluid or discrete particles.

In the first chapter of the book, we first introduce the multiphase flow phenomena and different flow regimes encountered in a typical two-phase flow system and introduce the simulation and modelling methods applicable to each regime. The outline of the book will also become clear in the introduction.

In Chap. 3, probability theory will be discussed and both single variable and multivariate probability density functions will be defined with some useful standard PDFs and their properties that will be used in the later chapters. Then a stochastic process will be defined using physical examples and numerical simulation to clarify the concept. Standard processes are introduced and their properties are examined using numerical simulations and analytical proofs. Readers already familiar with these concepts can safely skip this chapter and start at Chap. 3.

Kinetic and granular theory will be discussed to introduce the concept of the PDF transport equations and averaging process. This problem holds many resemblances to our particle-fluid system and the physical interpretation of many mathematical definitions will become clear in this chapter. Then we turn to the main problem and give the mathematical description of the fluid-particle system using both Eulerian and Lagrangian approaches and define the Mass Density Function, which is usually used to write the equations instead of a PDF due to its intuitive physical interpretation. The state vector of the system will be defined and transition from a deterministic definition to a stochastic definition, its necessity for a turbulent system and its consequence will be discussed and the derivation of the general transport equation will be detailed.

Using the general transport equations, the field equations are derived and different terms will be explained and interpreted physically. Application of the method to mono-sized and poly-sized particles will be introduced, and the consequences of adding a length scale to the state vector due to poly-dispersity of the flow will become clear. Finally, the closure problem will be discussed and several applicable methods will be introduced and suggestions will be provided.

Southampton, March 2014                                      J. S. Shrimpton
                                                              S. Haeri
                                                              Stephen J. Scott

# Contents

# Acronyms

| | |
|---|---|
| BBGKY | Bogoliubov Born Green Kirkwood Yvon |
| CK | Chapman–Kolmogorov |
| DFMM | Direct Fractional Method of Moments |
| DNS | Direct Numerical Simulation |
| EE | Eulerian–Eulerian |
| EL | Eulerian–Lagrangian |
| GL | Grunwald–Letnikov |
| KTGF | Kinetic Theory of Granular Flow |
| KTGK | Kinetic Theory of Gases |
| LES | Large Eddy Simulation |
| LPM | Laguerre Polynomial Method |
| MDF | Mass Density Function |
| MEM | Maximum Entropy Method |
| MGF | Moment Generating Function |
| NS | Navier–Stokes |
| ODE | Ordinary Differential Equation |
| PDF | Probability Density Function |
| RANS | Reynolds Average Navier–Stokes |
| RL | Rienmann–Liouville |
| SDE | Stochastic Differential Equation |

# Chapter 1
# Introduction

Multi-phase flows are encountered in many engineering and environmental systems. For example, controlling combustion to enable efficient fuel consumption, is only possible by understanding the atomization, dispersion, and evaporation processes of fuel droplets in the combustion system. The inverse problem, from an environmental perspective, could be condensation and collision of drops in clouds to produce rain. Particle separation in cyclones and filters, hydraulic conveying, liquid–solid separation, particle dispersion in stirred vessels, spray drying and cooling, mixing of immiscible liquids, liquid–liquid extraction, bubble columns, aeration of sewage water, and flotation are only some of the industrial processes involving a continuous and dispersed phase. Furthermore, in most processes, the continuous phase is turbulent.

There are many parameters, such as coalescence and breakup of bubbles [1], turbulence modulation [2], clustering and preferential accumulation [3, 4], that affect the efficiency of processes involving particulate flows. In some processes, increasing a flow parameter might be of the main concern, while in others a minimization of the same parameter might be essential. For instance, in internal combustion engines, a large interparticle separation implies large fuel evaporation rate and better fuel-air mixing and hence results in less soot production and more efficient combustion [5, 6]. For floating or separation processes, a small interparticle distance is needed to alleviate the clustering and separation of particles. There are many other complicated phenomena that occur in such systems involving interphase heat, mass and momentum transfer, which are not yet fully understood. Theoretically, all these interphase phenomena can be added to a simulation, but the presence of a very large range of scales in multi-phase and turbulent flow systems makes simulation extremely challenging.

In addition to the aforementioned scale problem, the field of multi-phase flow is extremely wide and can be classified and studied from many perspectives, as a result there are many different simulation techniques available, each applicable to specific subclasses and phenomena. Therefore, a brief discussion on the different types of the multi-phase flows is provided in this chapter which is, of course, by no means an exhaustive review of the literature.

J. S. Shrimpton et al., *Statistical Treatment of Turbulent
Polydisperse Particle Systems*, Green Energy and Technology,
DOI: 10.1007/978-1-4471-6344-2_1, © Springer-Verlag London 2014

**Fig. 1.1** Gas–solid flow
regimes—different regimes
observed by increasing gas
velocity from *left* to *right* [7]

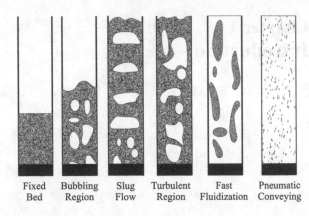

| Fixed | Bubbling | Slug | Turbulent | Fast | Pneumatic |
| Bed | Region | Flow | Region | Fluidization | Conveying |

## 1.1 Multi-phase Flows

The term "two-phase flow" refers to the situation where a mixture is formed either
by two immiscible fluids or a fluid and a solid phase. If the mixture consists of a fluid
and a solid phase, the solid phase will usually appear as small particles suspended in
the fluid phase; in this case, the fluid phase is connected continuously and is usually
referred to as carrier phase and the solid phase as dispersed. The dispersed phase can
be further classified into two different classes: mono- and poly-dispersed depending
on the size distribution of particles. Different flow regimes can be categorized by
considering a dense particulate phase, for instance, a fluidized bed. In this respect
Fig. 1.1 in a fluidized bed, adapted from Grace [7] shows the different regimes for a
gas–solid suspension.

The properties of the gas–solid regimes can be summarized as follows [7–10]:

- Fixed bed regime ($0 < U < U_{mf}$): Particles are quiescent; pressure drop increases
  by increasing velocity.
- Bubbling Fluidization ($U_{mf} < U < U_{ms}$): Voids form near the distributor and grow
  by coalescence, large irregular pressure fluctuations and well-defined surface.
- Slugging Fluidization ($U_{ms} < U < U_c$): Voids fill the most of the cross section,
  rise, and collapse of the top surface with regular frequency, large and regular
  pressure fluctuation.
- Turbulent Fluidization ($U_c < U < U_{se}$): Small voids and particle clusters are
  formed and destroyed, small pressure fluctuations and hard to distinguish the upper
  surface.
- Fast Fluidization ($U_{se} < U$): No upper surface, cluster of particles moving down
  specially near walls and dilute mixture removes particles in the middle.

Here, $U_{mf}$ is the minimum fluidization velocity, which is extensively studied,
and a number of equations are available [7, 8] for its calculation. $U_c$ is the crit-
ical superficial velocity at which the standard deviation of pressure fluctuations
passes through a maximum, and transition to turbulence fluidization occurs [11, 12].

**Fig. 1.2** Gas–liquid flow regimes in *vertical* tubes—different regimes observed by increasing gas velocity from *left* to *right* at constant liquid velocity

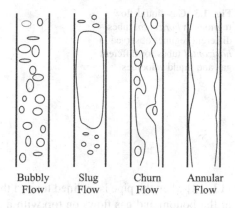

| Bubbly | Slug | Churn | Annular |
| Flow | Flow | Flow | Flow |

$U_{se}$ is the velocity at which the entrainment process starts where particles can no longer be maintained in the column unless entrained particles are captured and returned to the bed efficiently. In contrast, for columns with large diameters, or small particle diameters, $D_p$, the slugging regime may be bypassed altogether, for which equations can be found in [11].

Fluid–solid regimes are of little practical importance in horizontal, vertical, and inclined tubes vary, and therefore, the classification of the flow regimes is far more complicated. However, generally a fluid–fluid mixture falls into a variety of different regimes depending mainly on the relative superficial velocity of each phase [13, 14]. Figure 1.2 shows the different regimes observed in a vertical tube. The main flow regimes for fluid–fluid mixtures in vertical tubes and their characteristics can be summarized as follows [15, 16]:

- Bubbly flow: Discrete small bubbles are contained within the liquid continuous phase.
- Slug flow: Contains large "Taylor bubbles" with an ellipsoidal nose with diameters approaching the channel width; the gas is separated from the side walls by slowly descending liquid films.
- Churn or froth flow: It is formed by the breakdown of large gas bubbles in the slug flow with oscillatory or time varying characteristic.
- Annular flow: A liquid film forms at the channel side wall with a continuous gas core with large amplitude coherent waves usually present on the surface of the film.
- Mist flow: Discrete liquid drops are contained within the continuous gas phase which has specific applications in spray processes.

Note that in gas–liquid flows the transitions between different regimes are a function of both liquid and gas superficial velocities ($U_{sl}$, and $U_{sg}$) [17] thus, it can be said that in a constant superficial liquid velocity, $U_{sl}$, the aforementioned patterns are orderly observed by increasing the gas superficial velocity $U_{sg}$.

In addition to gas–liquid flow patterns mentioned earlier, there is also the possibility of forming stratified or wavy-stratified flow at low gas and liquid velocities

**Fig. 1.3** Gas–liquid flow regimes in *horizontal* tubes— different regimes observed in *horizontal* tubes for different gas and liquid velocities

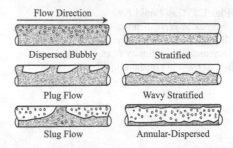

[13, 18] when the pipe is inclined toward the horizontal. In this regime, liquid flows at the bottom and gas flows on top with a quiescent well-defined boundary, which can become wavy with a slight increase in gas velocity [19], see Fig. 1.3. Flow patterns in inclined tubes are also studied extensively for downward, upward, and also for co- and counter-current flow conditions [20–23]. An exhaustive review of these regimes is outside the scope of current book. The present contribution is restricted to where one fluid appears as a continuous phase and the other phase as discrete and separate inclusions. This usually happens at small volume fraction of the dispersed phase.

## 1.2 Multi-phase Solution Strategies

Industrial processes involving fluid flow are usually accompanied by heat and mass transfer, and computational simulation techniques have increasingly been used for modeling and optimization purposes over the past two decades. These techniques can be classified in terms of the resolution of the interface between the phases and also the range of time scales, as follows:

- For engineering purposes, two approaches based on the Reynolds averaged NS equations are commonly applied, namely the two-fluid or Eulerian–Eulerian (EE) approach, see Fig. 1.4a and the Eulerian–Lagrangian (EL) methods, Fig. 1.4b. In these methods, the inter-phase interactions for momentum, heat, and mass transfer have to be extended by appropriate source/sink terms.

  In the EE (or two-fluid) approach, both phases are considered as interacting continua. Hence, properties such as the mass of particles per unit volume are considered as a continuous property, particle phase velocity, interfacial transfer of mass, momentum, or energy require averaging over the computational cells.

  The Eulerian–Lagrangian approach is only applicable to dispersed two-phase flows where the discrete nature of the individual particles are considered. In this method, particles are tracked in a Lagrangian manner by using a point-particle assumption and appropriate models are employed for the various forces acting on the particles. However, since the number of particles in a real simulation is usually too

**Fig. 1.4** Different simulation
techniques used for particle
flow systems. More modeling
is required to scale up the
simulations

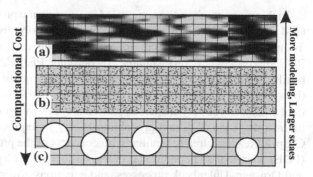

large to be tracked individually, another layer of simplification is usually added by
introducing "computational" particles or "parcels", which represent a number of real
particles with the same properties (i.e. size, velocity and temperature). Then, instead
of individual particles, the computational particles are tracked and properties such
as dispersed phase density and velocity are obtained by ensemble averaging [5].

- Direct numerical simulations (DNS) for the fluid phase, and a point-particle
  assumption for the particle phase. In this approach, finite dimensions of the particle are not considered and a Lagrangian approach is used to track the dispersed
  phase. This implies that a large number of particles may simultaneously be tracked
  through the computed time-dependent flow field by considering the relevant forces
  and where the effects of the particles on the flow field may or may not be considered. This approach is usually used for fundamental turbulent research in order to
  analyze the particle behavior in turbulent flows. Conventionally, pseudo-spectral
  methods pioneered by [24] and later extended by [25] are used to simulate turbulent
  flow. The literature is rich in one- and two-way coupling techniques, and the interested reader can consult [3, 26–32]. As a more practical alternative pseudo-spectral
  DNS methods, large eddy simulations (LES) combined with the Lagrangian tracking of point-particles have also been applied to the study of basic phenomena, such
  as particle dispersion in turbulent flows, inter-particle collisions [33], and particle behavior in channel flows [34]. This type of simulation is demonstrated in
  Fig. 1.4b.
- Direct numerical simulations of particulate flows resolving the flow around the
  particles to account for their finite dimensions have become feasible recently by
  virtue of the drastic increase of computational power. In such an approach, the
  time-dependent solution of the three-dimensional Navier-Stokes (NS) equations
  on a grid, which resolves the particles is obtained using the appropriate boundary
  conditions on the surfaces of the particles. Two main approaches that are used
  to implement the boundary conditions and resolve the particles are as follows:
  (i) body conformal mesh techniques and (ii) fixed mesh methods. The interested
  reader can consult [35–37] and the references therein for further information. These
  simulations are the most expensive of the techniques discussed in this section and
  are only currently feasible for small $[(O(10^3)]$ number of particles, see Fig. 1.4c.

In this monograph, we are interested in a systematic procedure for deriving the EE field equations of the fluid and particle phases for poly-dispersed systems. Chapter 3 discusses the various different approaches to this problem and demonstrates that PDF approaches are very suitable for deriving EE field equations. Mathematical results from the theory of stochastic differential equations (SDE) and probability theory will be applied extensively in this monograph to systematically derive EE field equations for poly-dispersed particulate flow systems. To this end, a review of the results of the SDE and probability theory is required, which is the subject of Chap. 2. Readers already familiar with concepts of the probability density functions (PDE), marginal and conditional PDFs, basic stochastic processes, such as Wiener and Ornstein–Uhlenbeck processes, and equations governing the time evolution of a PDF, i.e. Fokker–Planck equations, can skip Chap. 2.

# References

1. Gorokhovski M, Herrmann M (2008) Modeling primary atomization. Annu Rev Fluid Mech 40:343–366
2. Crowe C (2000) On models for turbulence modulation in fluid/particle flows. Int J Multiphase Flow 26:719–727
3. Scott SJ, Karnik AU, Shrimpton JS (2009) On the quantification of preferential accumulation. Int J Heat Fluid Flow 30:789–795
4. Shaw R (2003) Particle-turbulence interactions in atmospheric clouds. Annu Rev Fluid Mech 35:183–227
5. Shrimpton J (2003) Pulsed charged sprays: application ro disi engines during early injection. Int J Numer Meth Eng 58:513–536
6. Feath G (1983) Evaporation and combustion in sprays. IEEE Trans 1A–19:754–758
7. Grace J (1986) Contacting modes and behaviour classification of gas-solid and other two-phase suspensions. Can J of Chem Eng 64:1953–1966
8. Gidaspow (1994) Multiphase flow and fluidization. Academic Press, New York
9. Geldart D (1973) Types of gas fluidization. Powder Technol 7:185–195
10. Clift R, Grace J, Weber M (1978) Bubbles, drops and particles. Academic Press, New York
11. Bi HT, Grace JR (1995) Flow regime diagrams for gas-solid fluidization and upward transport. Int J Multiphase Flow 21:1229–1236
12. Bi HT, Grace JR, Lim KS (1995) Transition from bubbling to turbulent fluidization. Ind Eng Chem Res 34:4003–4008
13. Mandhane M, Gregory GA, Aziz K (1974) A flow pattern map for gas-liquid flow in horizontal pipes. Int J Multiphase Flow 1:537–553
14. Beggs D, Brill J (1973) A study of two-phase flow in inclined pipes. J Petrol Technol 25:607–617
15. Xu J, Cheng P, Zhao T (1999) Gas liquid two-phase flow regimes in rectangular channels with mini/micro gaps. Int J Multiphase Flow 25:411–432
16. Furukawa T, Fukano T (2001) Effect of lquid viscosity on flow patterns in vertical upward gas-liquid two-phase flow. Int J Multiphase Flow 27:1109–1126
17. Taitel Y, Bornea D, Dukler AE (1980) Modelling flow pattern transitions for steady upward gas-liquid flow in vertical tubes. AIChE J 26:345–354
18. Barnea D, Shoham O, Taitel Y, Dukler AE (1980) Flow pattern transition for gas-liquid flow in horizontal and inclined pipes. Int J Multiphase Flow 6:217–225
19. Liu Y, Yang W, Wang J (2008) Experimental study for the stratified to slug flow regime transition mechanism of gas-oil two-phase flow in horizontal pipe. Frontiers Energy Power Eng China 2:152–157

20. Pantzali M, Mouza A, Paras S (2008) Counter-current gas-liquid flow and incipient flooding in inclined small diameter tubes. Chem Eng Sci 63:3966–3978
21. Dyment A, Boudlal A (2004) A theoretical model for gas-liquid slug flow in down inclined ducts. Int J Multiphase Flow 30:521–550
22. Zapke A, Krger DG (2000) Countercurrent gas-liquid flow in inclined and vertical ducts-i: flow patterns, pressure drop characteristics and flooding. Int J Multiphase Flow 26:1439–1455
23. Grolman E, Fortuin JMH (1997) Gas-liquid flow in slightly inclined pipes. Chem Eng Sci 52:4461–4471
24. Riley JJ, Patterson GS (1974) Diffusion experiements with numerically integrated isotropic turbulence. Phys Fluids 17:292–297
25. Rogallo R (1981) Numerical experiments in homogeneous turbulence. Technical Report 81315, NASA
26. Squires K, Eaton J (1990) Preferential concentration of particles by turbulence. Phys Fluids 3:1169–1178
27. Squires K, Eaton J (1991) Measurements of particle dispersion obtained from direct numerical simulations of isotropic turbulence. J Fluid Mech 226:1–31
28. Squires K, Eaton J (1991) Lagrangian and eulerian statistics obtained from direct numerical simulations of homogeneous turbulence. Phys Fluids 3:130–143
29. Eaton J (2009) Two-way coupled turbulence simulations of gas-particle flows using point-particle tracking. Int J Multiphase Flow 35:792–800
30. Elghobashi S, Truesdell G (1992) Direct simulation of particle dispersion in decaying isotropic turbulence. J Fluid Mech 242:655–700
31. Elghobashi S, Truesdell G (1993) On the two-way interaction between homogeneous turbulence and dispersed solid particles part 1: turbulence modification. Phys Fluids A5:1790–1801
32. Boivin M, Simonin O, Squires KD (1998) Direct numerical simulation of turbulence modulation by particles in isotropic turbulence. J Fluid Mech 375:235–263
33. Lavieville J, Deutsch E, Simonin O (1995) Large eddy simulation of interactions between colliding particles and a homogeneous isotropic turbulence field. gas-particle flows. ASME 228:359–369
34. Wang Q, Squires K, Chen H, McLaughlin J (1997) On the role of the lift force in turbulence simulations of particle deposition. Int J Multiphase Flow 23:749–763
35. Haeri S, Shrimpton J (2012) On the application of immersed boundary, fictitious domain and body-conformal mesh methods to many particle multiphase flows. Int J Multiphase Flow 40:38–55
36. Haeri S, Shrimpton J (2013) A correlation for the calculation of the local nusselt number around circular cylinders in the range $10 \leq re \leq 250$ and $0.1 \leq pr \leq 40$. Int J Heat Mass Transfer 59:219–229
37. Haeri S, Shrimpton J (2013) A new implicit fictitious domain method for the simulation of flow in complex geometries with heat transfer. J Comput Phys 237:21–45

# Chapter 2
# PDF Method: A Stochastic Framework

To model the desired macroscopic quantities, the easiest way is to write closed PDEs for specific quantities. In this process, additional unknowns are added to the system and the success of this method depends on the accuracy with which the unclosed terms explicitly can be based on macroscopic laws or using a model which does not depart dramatically from reality. As discussed in Chap. 1, the applicability of this method to complicated multiphase flows is limited because the average of a complicated function of a variable such as $\langle f(\varphi) \rangle$ must be defined based on the available information, which is typically limited to the first- and second-order moments of the variable $\varphi$, i.e. $\langle \varphi \rangle$ and $\langle \varphi^2 \rangle$. For complicated reactive or poly-dispersed flows, this information is typically insufficient to accurately define such functions. A full or direct numerical simulation is also not feasible for complicated flows and geometries, and the only reasonable solution would be what is usually referred to as a mesoscopic approach.

In this monograph, we mainly focus on the mesoscopic approach based on PDFs, and hence, a brief introduction to the existing stochastic and statistical methods in a form suitable for modelling poly-dispersed turbulent particulate flows is required and is provided in this chapter. The material in this chapter is presented without formal proof, and the interested reader can consult [1–4] for further information. Several examples are, however, provided to demonstrate the significance of different stochastic processes.

## 2.1 Definition of a Stochastic Process

A stochastic variable, $X$, in applied sciences is usually defined directly from its probability density function (PDF). For now, assume that this random (stochastic) variable, $X$, is a scalar, e.g. one-dimensional velocity in Brownian motion or temperature of a particle. This stochastic variable, $X$, can take a range of possible values

J. S. Shrimpton et al., *Statistical Treatment of Turbulent Polydisperse Particle Systems*, Green Energy and Technology, DOI: 10.1007/978-1-4471-6344-2_2, © Springer-Verlag London 2014

$x \in S$ where $S$ can be a discrete or continuous set such as $\mathbb{R}$ or $\mathbb{R}^d$ and the probability that $X$ takes the values between $x$ and $x + dx$ is

$$\Pr(x < X < x + dx) = \mathscr{P}(x)dx. \tag{2.1}$$

In the same fashion, we can define a multivariate distribution, also called joint probability distribution of $r$ variables $X_1, \ldots, X_r$. Taking a subset $s$ of $r$, $(s < r)$, marginal distribution of $s$ variables is the probability that $X_1, \ldots, X_s$ take the values $x_1, \ldots, x_s$, regardless of values of $X_{s+1}, \ldots, X_r$, i.e.

$$\mathscr{P}(x_1, \ldots, x_s) = \int \mathscr{P}(x_1, \ldots, x_s, x_{s+1}, \ldots, x_r)dx_{s+1}, \ldots, dx_r \tag{2.2}$$

It is also possible for one to assign fixed values to $X_{s+1}, \ldots, X_r$ and ask for the joint probability of the remaining variables which is called conditional probability of $X_1, \ldots, X_s$ and defined by $\mathscr{P}(x_1, \ldots, x_s | x_{s+1}, \ldots, x_r)$. The joint probability of $X_1, \ldots, X_r$ is equal to the marginal probability of $X_{s+1}, \ldots, X_r$, to have the values $x_{s+1}, \ldots, x_r$, multiplied by the conditional probability of $X_1, \ldots, X_s$, given the values of $X_{s+1}, \ldots, X_r$:

$$\mathscr{P}(x_1, \ldots, x_r) = \mathscr{P}(x_{s+1}, \ldots, x_r)\mathscr{P}(x_1, \ldots, x_s | x_{s+1}, \ldots, x_r) \tag{2.3}$$

Once the stochastic variable $X$ is defined, other stochastic variables, namely $Y$, can be derived from it by a mapping, $f$. These new variables can also be functions of time, $t$, and we can write

$$Y_X(t) = f(X, t) \tag{2.4}$$

By replacing $X$ by $x$, one possible value, we get $Y_x(t) = f(x, t)$, an ordinary function called a realization of the process. Therefore, the stochastic process in physical sense is an ensemble of these realizations. By measuring the values, $x_0, \ldots, x_n$ at times $t_0, \ldots, t_n$ where $t_0 \leq t_n$, we can completely describe the process by the joint probability density function $\mathscr{P}(x_n, t_n; \ldots; x_0, t_0)$. In case of complete independence, we can write

$$\mathscr{P}(x_n, t_n; \ldots; x_0, t_0) = \prod_i \mathscr{P}(x_i, t_i) \tag{2.5}$$

This means that the values of $X$ at time $t$ are independent of its values in the past or future. Also note that the $x_i$ values could each be a vector, e.g. velocity components or the whole phase space vector, i.e. all the velocity and position components in addition to other scalars. We use bold symbols such as $\mathbf{X}$ or $\mathbf{Z}$, and their corresponding values $\mathbf{x}$ and $\mathbf{z}$, to indicate that these stochastic variables are in fact vectors, which are also functions of time, defining the whole phase space.

## 2.2 Markov Process

The general process defined by[1]:

$$\mathscr{P}(\mathbf{x}_n, t_n; \ldots; \mathbf{x}_0, t_0) \tag{2.6}$$

is very difficult to handle since one needs the knowledge of all the previous points in time to describe such a process. Accordingly, the problem is usually restricted to a family of processes known as Markov processes where the knowledge of present state of the system completely describes the whole process. In other words, the past history of process has no effect on the future evolution of the process or mathematically

$$\mathscr{P}(\mathbf{x}_n, t_n | \mathbf{x}_{n-1}, t_{n-1}; \ldots; \mathbf{x}_0, t_0) = \mathscr{P}(\mathbf{x}_n, t_n | \mathbf{x}_{n-1}, t_{n-1}) \tag{2.7}$$

In Eq. (2.7), $\mathscr{P}(\mathbf{x}_n, t_n | \mathbf{x}_{n-1}, t_{n-1})$ is also called the transition probability. This assumption is very powerful and means everything can be defined in terms of a transition probability $\mathscr{P}(\mathbf{x}_n, t_n | \mathbf{x}_{n-1}, t_{n-1})$ and an initial probability $\mathscr{P}(\mathbf{x}_0, t_0)$. Thus, for example,

$$\mathscr{P}(\mathbf{x}_2, t_2; \mathbf{x}_1, t_1; \mathbf{x}_0, t_0) = \mathscr{P}(\mathbf{x}_2, t_2 | \mathbf{x}_1, t_1) \mathscr{P}(\mathbf{x}_1, t_1 | \mathbf{x}_0, t_0) \mathscr{P}(\mathbf{x}_0, t_0) \tag{2.8}$$

Note that the process can be continuous or discontinuous, regardless of the nature of the variable $\mathbf{X}$. For example, the sample space of classical Brownian motion of a particle immersed in a collection of light molecules, with assumption of hard sphere collisions, by virtue of instant jumps, is not continuous in spite of the fact that the range of velocities is continuous.

## 2.3 The Chapman–Kolmogorov Equation

For a general stochastic process, we can write:

$$\mathscr{P}(\mathbf{x}_2, t_2 | \mathbf{x}_0, t_0) = \int \mathscr{P}(\mathbf{x}_2, t_2 | \mathbf{x}_1, t_1; \mathbf{x}_0, t_0) \mathscr{P}(\mathbf{x}_1, t_1 | \mathbf{x}_0, t_0) \mathrm{d}\mathbf{x}_1 \tag{2.9}$$

Introducing the Markov property, i.e. Eq. (2.7) into Eq. (2.9), results in the celebrated Chapman–Kolmogorov (CK) equation:

$$\mathscr{P}(\mathbf{x}_2, t_2 | \mathbf{x}_0, t_0) = \int \mathscr{P}(\mathbf{x}_2, t_2 | \mathbf{x}_1, t_1) \mathscr{P}(\mathbf{x}_1, t_1 | \mathbf{x}_0, t_0) \mathrm{d}\mathbf{x}_1 \tag{2.10}$$

---

[1] In this chapter, we intend to use the tensor notation for all variables except when a variable is used as an argument of a function, e.g. $\mathscr{P}(\mathbf{x})$, and when used as an integration variable, where using the tensor notation is confusing and leads to misinterpretation of the equation.

Equation (2.10) simply states that the probability of a process ending in state $(\mathbf{x}_2, t_2)$ given the initial state $(\mathbf{x}_0, t_0)$ is equal to the sum of all possible paths from $(\mathbf{x}_0, t_0) \rightarrow (\mathbf{x}_2, t_2)$. The CK equation is a complex nonlinear equation relating all conditional probabilities to each other. Therefore, it is convenient to derive a differential form of CK equation which is easier to handle and physically easier to interpret. For example, the differential form can be derived based on a trajectory point of view [2].

Now considering the time evolution of the expectation of a twice differentiable function and using the CK equation, Eq. (2.10), the differential CK equation [2, 4, 5] can be derived as

$$\frac{\partial \mathscr{P}(\mathbf{x}, t | \mathbf{x}_0, t_0)}{\partial t} = -\sum_i \frac{\partial}{\partial x_i} \left( A_i(\mathbf{x}, t) \mathscr{P}(\mathbf{x}, t | \mathbf{x}_0, t_0) \right)$$

$$+ \sum_{ij} \frac{1}{2} \frac{\partial^2}{\partial x_i \partial x_j} \left( B_{ij}(\mathbf{x}, t) \mathscr{P}(\mathbf{x}, t | \mathbf{x}_0, t_0) \right)$$

$$+ \int \left( J(\mathbf{x} | \mathbf{z}, t) \mathscr{P}(\mathbf{z}, t | \mathbf{x}_0, t_0) - J(\mathbf{z} | \mathbf{x}, t) \mathscr{P}(\mathbf{x}, t | \mathbf{x}_0, t_0) \right) \mathrm{d}\mathbf{z}.$$

$$(2.11)$$

In the differential CK equation, Eq. (2.11), the first and second terms define the drift and diffusion processes, respectively, and the last integral defines a jump process. Jump processes will be discussed thoroughly in Sect. 2.6 and a pure drift process, which is also known as the deterministic process (or Liouville's Equation) and a diffusion process (or Fokker–Planck Equation), is discussed in Sect. 2.5. The differential CK equation simply states that having the initial distribution of some variable $\mathbf{x}$ at time $t_0$, the final distribution at any time in the future can be found using this equation.

We are sometimes interested in knowing the probability of a system at state $\mathbf{x}$ at time $t_0$ to finish in a subset of states $S$. In this case, we want to know for every state $\mathbf{x}$ at time $t < t_{\text{final}}$ what is the probability of ending up in the subset $S$. In this case, the differential CK equation can be restated as the backward CK equation:

$$\frac{\partial \mathscr{P}(\mathbf{x}_0, t_0 | \mathbf{x}, t)}{\partial t} = -\sum_i A_i(\mathbf{x}, t) \frac{\partial}{\partial x_i} \left( \mathscr{P}(\mathbf{x}_0, t_0 | \mathbf{x}, t) \right)$$

$$- \frac{1}{2} \sum_{ij} B_{ij}(\mathbf{x}, t) \frac{\partial^2}{\partial x_i \partial x_j} \left( \mathscr{P}(\mathbf{x}_0, t_0 | \mathbf{x}, t) \right)$$

$$+ \int \left( J(\mathbf{z} | \mathbf{x}, t) \mathscr{P}(\mathbf{x}_0, t_0 | \mathbf{x}, t) - J(\mathbf{z} | \mathbf{x}, t) \mathscr{P}(\mathbf{x}_0, t_0 | \mathbf{z}, t) \right) \mathrm{d}\mathbf{z}.$$

$$(2.12)$$

Next, we will define the most basic stochastic process (Wiener process) and discuss the solution of forward CK equations for this simple case, and in Sect. 2.7, we provide a solution to the backward equation.

## 2.4 Wiener Process

The Wiener process is the most fundamental of continuous time stochastic processes and is essentially a diffusion process with $A_i = 0$, $B_{ij} = 1$ and $J = 0$. A standard Wiener process on the interval $[0, T]$ is a random variable $W(t)$ that depends continuously on $t \in [0, T]$ and satisfies the following conditions: (i) the increments of the Wiener process are distributed according to a normal (Gaussian) distribution; (ii) trajectories of $W(t)$ are continuous but cannot be differentiated; (iii) Increments of $W(t)$ are stationary and independent; and (iv) trajectories are of unbounded variation in finite time intervals. Numerical solution to the SDEs is out of the scope of this book and can be found elsewhere [5]. However, simulating a Wiener process is straightforward by using the property (i), and one only needs to sample from a normal distribution and add this to the current state of the system using the relation $W_{i+1} = W_i + dW_i$ and scale the results. The Wiener process physically translates to a pure Brownian motion without any frictional coefficient defined by [6, 7]

$$\frac{dU_{p,x}}{dt} = W_{p,x}(t), \tag{2.13}$$

with initial condition:

$$U_{p,x}|_{t=0} = U_{p,x0}, \tag{2.14}$$

where $W_{p,x}$ is a random rapidly fluctuating force per unit mass exerted on the particle $p$, due to collisions with other smaller particles. This type of differential equation with a stochastic function on one side is known as the Langevin equation, c.f. [1, 2]. The right-hand side of Eq. (2.13) and so the Langevin equation can include other terms, such as a friction ($\mu U_{p,x}$) term [6].

Figure 2.1 shows a sample path of such motion simulated by integrating Eq. (2.13) directly. The pure stochastic motion without any friction is shown with $\mu = 0$ in this figure. Evidently, the solution deviates from the initial conditions significantly. However, setting the frictional coefficient to $\mu = -1$ acts as a restoring force, and the solution only vibrates around the initial value which physically translates to a random motion but with a bounded molecular velocity and is a better model for the physical phenomenon.

The solution of the forward differential CK equation for the Wiener process is given in Fig. 2.2. For the forward equation, the proper initial condition is $\mathscr{P}(\mathbf{x}, t_0|\mathbf{x_0}, t_0) = \delta(\mathbf{x} - \mathbf{x_0})$, which simply means that we know the solution at $t = 0$ with probability one. Figure 2.3 shows the effect of the restoring force, which prevents the PDF from rapid evolution, and the value of the velocity remains around

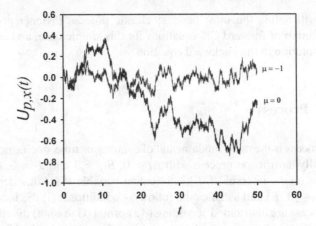

**Fig. 2.1** 1D Brownian motion—two sample path of the Brownian motion with and without friction coefficient $\mu$. Friction force acts as a restoring force that keeps the velocity deviations bounded

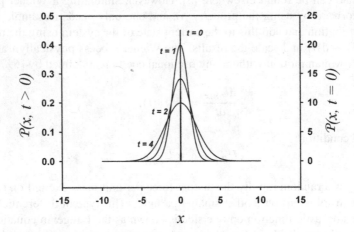

**Fig. 2.2** Solution of the forward CK equation for a Wiener process—the initial condition is known with probability one and thus is a scaled delta function at $t = 0$, and the PDF then evolves and becomes flatter in time, increasing the uncertainty in the solution

the initial conditions with probability one. This process with a linear drift added to the Wiener process is also known as Ornstein–Uhlenbeck process. Figure 2.4 shows the changes in the standard deviation, $\sigma$, of the PDF in time: smaller standard deviation corresponds to the smaller probability of occurrence of velocities far from the mean, while large values show that the extreme velocities are more probable. This simple example clarifies the connection between the trajectory and PDF point of view and also shows why the trajectory point of view resolves more information than the PDF point of view.

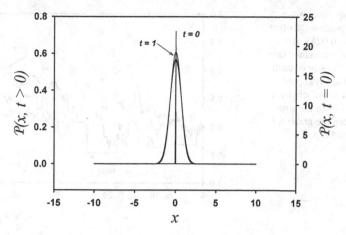

**Fig. 2.3** Solution of the forward CK equation for the Brownian motion—a Brownian motion with $\mu = -1$. Due to the certainty of the initial condition, it takes the form of the delta function. In this case, restoring force keeps the solution around a mean and the PDF remains constant for $t > 2$

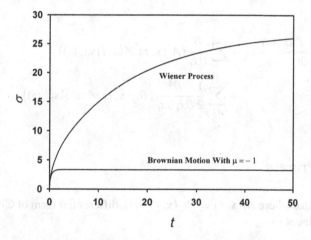

**Fig. 2.4** Comparison between standard deviations—changes in standard deviation of the PDF of the Wiener process and Brownian motion with $\mu = -1$ for $t \in [0, 50]$. It increases significantly for the Wiener process, while it rapidly reaches a small constant value for the Brownian motion with $\mu = -1$

## 2.5 Diffusion Process

A diffusion process is another subset of Markov process applicable to many physical systems where the sample path is continuous. In this case, $J$ should be zero in the differential CK equation, Eq. (2.11), as this term represents discontinuities, thereby reducing the complexity of the equation describing this process. Thus, the final form

**Fig. 2.5** Jump process—a
typical jump process (labelled
by $\text{Pr}_{accept} = 0.025$). By
increasing the acceptance rate
and reducing the size of each
jump, we can always approximate a diffusion process by a
jump process. Note the similarities between the graph for
$\text{Pr}_{accept} = 1$ and Fig. 2.1

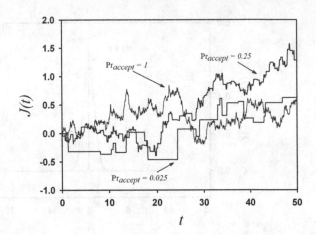

of the equation, which is commonly referred to as the Fokker–Planck equation,
becomes

$$\frac{\partial \mathscr{P}(\mathbf{x}, t | \mathbf{x}_0, t_0)}{\partial t} = -\sum_i \frac{\partial}{\partial x_i} \left( A_i(\mathbf{x}, t) \mathscr{P}(\mathbf{x}, t | \mathbf{x}_0, t_0) \right)$$

$$+ \sum_{ij} \frac{1}{2} \frac{\partial^2}{\partial x_i \partial x_i} \left( B_{ij}(\mathbf{x}, t) \mathscr{P}(\mathbf{x}, t | \mathbf{x}_0, t_0) \right). \qquad (2.15)$$

## 2.6 Jump Process

Consider the case where $A_i(\mathbf{x}, t) = B_{ij}(\mathbf{x}, t) = 0$, differential form of CK equation,
Eq. (2.11), reduces to

$$\frac{\partial \mathscr{P}(\mathbf{x}, t | \mathbf{x}_0, t_0)}{\partial t} = \int \left( J(\mathbf{x} | \mathbf{z}, t) \mathscr{P}(\mathbf{z}, t | \mathbf{x}_0, t_0) - J(\mathbf{z} | \mathbf{x}, t) \mathscr{P}(\mathbf{x}, t | \mathbf{x}_0, t_0) \right) d\mathbf{z}.$$

$$(2.16)$$

Equation (2.16) is usually known as the master equation. It should be noted that
the master equation can also be interpreted as a simple gain–loss of probabilities such
that the first term in the integral is the gain due to transition from other states and the
second is the loss due to transition to other states. It can be said that every diffusion
process can be approximated by a jump process. This means that in the limit of
infinitely small jump sizes, the master equation becomes a Fokker–Planck equation.
Figure 2.5 shows the sample path of a jump process and this scaling assumption
(see also [2]).

The equivalence between the jump process and diffusion process can also be demonstrated analytically by deriving the Fokker–Planck equation from the master equation. To show this, without loss of generality, we assume a univariate stochastic process. Furthermore, we need to define the increments of a stochastic process. The increment defines the behaviour of the process over small time periods. The increment $\Delta_s$ of a general stochastic process $X(t)$ in a positive time interval $s$ is defined by

$$\Delta_s X(t) = X(t+s) - X(t), \quad \text{for} \ s > 0, \tag{2.17}$$

and

$$dX(t) = \lim_{s \to 0} [X(t+s) - X(t)] = \lim_{s \to 0} [\Delta_s X(t)]. \tag{2.18}$$

Here, $s$ is positive; hence, the increment is considered to be forward in time. A process can be considered to be the sum of increments.

$$X(t_N) = X(t_0) + \sum_{k=1}^{N} \Delta_{t_k - t_{k-1}} X(t_{k-1}). \tag{2.19}$$

The PDF of the increment $\Delta_s X(t)$ conditional on $X(t) = x$ is denoted as $g(\hat{x}; s, x, t)$ where $\hat{x}$ represents the phase space increment. If $s$ is taken to be $t_2 - t_1$, then $X(t_1)$ can be expressed as

$$X(t_1) = X(t_2) - \Delta_s X(t_1), \tag{2.20}$$

hence, we can rewrite

$$\mathscr{P}(x_2; t_1 + s | x_1, t_1) = g(\hat{x}; s, x_2 - \hat{x}, t_1). \tag{2.21}$$

Now, the CK equation (2.10) is rewritten

$$\mathscr{P}(x_2; t_1 + s | x_0, t_0) = \int g(\hat{x}; s, x_2 - \hat{x}, t_1) \mathscr{P}(x_2 - \hat{x}; t_1 | x_0, t_0) d\hat{x}, \tag{2.22}$$

or

$$\mathscr{P}(x; t_1 + s | x_0, t_0) = \int g(\hat{x}; s, x - \hat{x}, t_1) \mathscr{P}(x - \hat{x}; t_1 | x_0, t_0) d\hat{x}. \tag{2.23}$$

In the CK equation, both $g$ and $p$ on the right-hand side involve the argument $x - \hat{x}$. Using a Taylor series expansion,[2] we have

---

[2] Taylor series expansion of $f(x)$ about a point $x - a$ is given by

$$f(x) = f(a) + f'(a)(x-a) + \frac{f''(a)}{2!}(x-a)^2 + \frac{f^3(a)}{3!}(x-a)^3 + \cdots + \frac{f^n(a)}{n!}(x-a)^n + \cdots.$$

$$\mathscr{P}(x; t_1 + s|x_0, t_0) = \mathscr{P}(x_1; t_1|x_0, t_0)$$

$$+ \int \sum_{n=1}^{\infty} \frac{(-\hat{x})^n}{n!} \frac{\partial^n}{\partial x^n} \left[ g(\hat{x}; s, x_1, t_1) \mathscr{P}(x_1; t_1|x_0, t_0) \right] d\hat{x}.$$

(2.24)

Rearranging and dividing through by $s$ give

$$\frac{1}{s} \left[ \mathscr{P}(x; t_1 + s|x_0, t_0) - \mathscr{P}(x_1; t_1|x_0, t_0) \right]$$

$$= \frac{1}{s} \int \sum_{n=1}^{\infty} \frac{(-\hat{x})^n}{n!} \frac{\partial^n}{\partial x_n} \left[ g(\hat{x}; s, x_1, t_1) \mathscr{P}(x_1; t_1|x_0, t_0) \right] d\hat{x}, \qquad (2.25)$$

and taking the limit $s \to 0$ term by term gives

$$\lim_{s \to 0} \frac{1}{s} \left[ \mathscr{P}(x; t_1 + s|x_0, t_0) - \mathscr{P}(x_1; t_1|x_0, t_0) \right] = \frac{\partial}{\partial x} \mathscr{P}(x; t|x_0, t_0), \qquad (2.26)$$

for the left-hand side and

$$\lim_{s \to 0} \frac{1}{s} \int \sum_{n=1}^{\infty} \frac{(-\hat{x})^n}{n!} \frac{\partial^n}{\partial x_n} \left[ g(\hat{x}; s, x_1, t_1) \mathscr{P}(x_1; t_1|x_0, t_0) \right] d\hat{x}$$

$$= \sum_{n=1}^{\infty} \frac{(-1)^n}{n!} \frac{\partial^n}{\partial x_n} \left[ \lim_{s \to 0} \frac{1}{s} \left\{ \int \hat{x} g(\hat{x}; s, x_1, t_1) d\hat{x} \right\} \mathscr{P}(x_1; t_1|x_0, t_0) \right]$$

$$= \sum_{n=1}^{\infty} \frac{(-1)^n}{n!} \frac{\partial^n}{\partial x_n} \left[ B_n(x, t) \mathscr{P}(x_1; t_1|x_0, t_0) \right], \qquad (2.27)$$

for the right-hand side. Combining terms gives us the Kramers–Moyal equation

$$\frac{\partial}{\partial x} \mathscr{P}(x; t|x_0, t_0) = \sum_{n=1}^{\infty} \frac{(-1)^n}{n!} \frac{\partial^n}{\partial x_n} \left[ B_n(x, t) \mathscr{P}(x_1; t_1|x_0, t_0) \right]. \qquad (2.28)$$

At this stage, it is still assumed that the parameters $B_n$ exist for all $n > 0$. The Kramers–Moyal equation is valid for general stochastic processes. If $t > t_0$, then we have the initial condition

$$\mathscr{P}(x; t|x_0, t_0) = \delta(x - x_0), \qquad (2.29)$$

and for a diffusion process, we know that $B_n = 0$ for $n \geq 3$. Given these conditions, the Kramers–Moyal equation simplifies to give the Fokker–Planck equation

$$\frac{\partial}{\partial t}\mathscr{P}(x; t|x_0, t_0) = -\frac{\partial}{\partial x}[a(x,t)\mathscr{P}(x; t|x_0, t_0)] + \frac{1}{2}\frac{\partial^2}{\partial x^2}\Big[b(x,t)^2\mathscr{P}(x; t|x_0, t_0)\Big].$$

(2.30)

The Fokker–Planck equation describes the evolution of the transitional PDF $\mathscr{P}(x; t|x_0, t_0)$ for a stochastic diffusion process. To obtain the evolution equation of the marginal PDF $\mathscr{P}(x; t)$, simply multiply by $\mathscr{P}(x_0; t_0)$ and integrate over $x_0$ to give

$$\frac{\partial}{\partial x}\mathscr{P}(x; t) = -\frac{\partial}{\partial x}[a(x,t)\mathscr{P}(x,t)] + \frac{1}{2}\frac{\partial^2}{\partial x^2}\Big[b(x,t)^2\mathscr{P}(x,t)\Big]. \qquad (2.31)$$

This equation is for the stochastic process $X(t)$ with drift and diffusion coefficients $a(x; t)$ and $b(x; t)^2$, respectively.

## 2.7  Stochastic Differential Equations

A stochastic differential equation (SDE) is a differential equation in which one or more terms are stochastic processes resulting in a solution which itself is a stochastic process. A simple SDE would be that of a Brownian motion introduced in Sect. 2.4. Here, we restrict attention to the relations between SDEs and the Fokker–Planck equation and will not discuss the SDE theory in detail.

A general stochastic differential equation also known as Langevin equation has the form:

$$dZ_i(t) = A_i(\mathbf{Z}(t), t)dt + B_{ij}(\mathbf{Z}(t), t)dW_j(t), \qquad (2.32)$$

where $W_j$ are a set of independent Wiener processes. It is important to notice that both the drift vector $A_i$ and the diffusion matrix $B_{ij}$ are functions of state vector variables $Z_i(t)$. To get from this equation to the corresponding Fokker–Planck equation, one should consider the time development of an arbitrary $f(\mathbf{Z}(t))$, and using the rules of Itô calculus, it is easy to show [2, 3] that the corresponding Fokker–Planck equation in n-dimensional sample space is

$$\frac{\partial \mathscr{P}}{\partial t} = -\frac{\partial}{\partial z_i}[A_i(\mathbf{z}; t)\mathscr{P}] + \frac{1}{2}\frac{\partial^2}{\partial z_i \partial z_j}[(\mathbf{B}\mathbf{B}^T)_{ij}(\mathbf{z}; t)\mathscr{P}]. \qquad (2.33)$$

In the above equation, $\mathbf{B}^T$ is the transpose of $\mathbf{B}$. If the diffusion matrix $B_{ij} = 0$, then the SDE reduces to the deterministic process:

$$dZ_i(t) = A_i(\mathbf{Z}(t), t)dt \qquad (2.34)$$

With corresponding Liouville equation,

$$\frac{\partial \mathscr{P}}{\partial t} = -\frac{\partial}{\partial z_i}[A_i(\mathbf{z}; t)\mathscr{P}]. \tag{2.35}$$

This is a completely deterministic system, i.e. if $z_i(\mathbf{x}_0, t)$ is the solution to Eq. (2.34) with initial conditions $z_i(\mathbf{x}_0, t_0) = x_{i,0}$, then $\mathscr{P}(\mathbf{x}, t|\mathbf{x}_0, t_0) = \delta(\mathbf{x} - \mathbf{z}(\mathbf{x}_0, t))$ is the solution to Eq. (2.35) with initial conditions $\mathscr{P}(\mathbf{x}, t_0|\mathbf{x}_0, t_0) = \delta(\mathbf{x} - \mathbf{x}_0)$. The proof is best obtained by direct substitution of

$$\mathscr{P}(\mathbf{x}, t|\mathbf{x}_0, t_0) = \delta(\mathbf{x} - \mathbf{z}(\mathbf{x}_0, t)), \tag{2.36}$$

into the RHS of Eq. (2.35) (note that $x_i$ is the independent variables):

$$-\frac{\partial}{\partial x_i}[A_i(\mathbf{x}; t)\delta(\mathbf{x} - \mathbf{z}(\mathbf{x}_0, t)] = -\frac{\partial}{\partial x_i}[A_i(\mathbf{z}; t)\delta(\mathbf{x} - \mathbf{z}(\mathbf{x}_0, t)]$$

$$= -A_i(\mathbf{z}; t)\frac{\partial}{\partial x_i}[\delta(\mathbf{x} - \mathbf{z}(\mathbf{x}_0, t)]. \tag{2.37}$$

The property of delta function is used in moving from first to second line and the fact that $A_i$ is not a function of $\mathbf{x}$ anymore, to move from second to third line. Also substituting Eq. (2.36) into the LHS of Eq. (2.35) and using the chain rule and again using the properties of the delta function give

$$\frac{\partial}{\partial t}[\delta(\mathbf{x} - \mathbf{z}(\mathbf{x}_0, t))] = -\frac{\partial}{\partial x_i}[\delta(\mathbf{x} - \mathbf{z}(\mathbf{x}_0, t))]\frac{dz_i(t)}{dt}. \tag{2.38}$$

Now using $dZ_i(t) = A_i(\mathbf{Z}(t), t)dt$ in Eq. (2.38) and changing to phase space notation Eq. (2.37) are restored.

Using an approach based on the Fokker–Planck equation provides a unique opportunity to add many physical phenomena to the model very conveniently, and also, information about the model can be inferred at this stage. For example, it is well known that inhomogeneous turbulence results in non-Gaussian PDFs [8] and this behaviour can easily be included in the model using a nonlinear drift term.

Figure 2.6 shows the evolution of the PDF of a SDE derived by adding the nonlinear drift $\sin(X_t)$ to a Wiener process. An appropriate final condition for the backward CK equation can be the box function $\mathbb{1}_{[-5,5]}$. The solution shows that the non-Gaussian behaviour is easily captured using this nonlinear drift term, and the probability of a solution starting at any previous time ending up in state $\mathbb{1}_{[-5,5]}$ is the solution to the backward equation.

**Fig. 2.6** Solution of the backward CK equation—a nonlinear drift term $\sin(X_t)$ produces a non-Gaussian behaviour of the PDF showing the benefits of modelling at a mesoscopic level

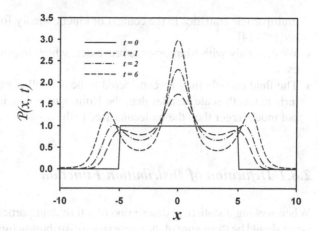

## 2.8 Fluid–Particle Systems

In this section, we show how the results of the probability theory and stochastic calculus presented in the previous section can be used to derive the various equations for modelling poly-dispersed particulate flows. Characterization of individual particles (solid or fluid) requires detailed knowledge of particle position, velocity, size, shape, rotation and temperature at a specified time. The equations derived in this manner are not ordinary PDEs, but are PDEs in a hyperspace of several dimensions, and this makes the direct solution to these equations a formidable task [9]. Therefore, it is necessary to reduce the dimensionality of the problem and make some simplifying assumptions to reduce the solution cost and also the model development process. We assume throughout the rest of this review that

- All particles are spherical, and hence, particles are only characterized by their diameter.
- The state vector of particle $i$ consists of its position, velocity and diameter $(\mathbf{X}^{(i)}, \mathbf{U}^{(i)}, \Phi^{(i)})$.
- Particles are non-interacting and can be treated as point processes, which implies a particle with no spatial extent. In Sect. 4.2, we discussed that here, we are interested in turbulence–particle interaction, and thus, particle–particle interactions are discarded in the following discussion. However, it should be obvious at this point that these forces can be included as a jump process in our general framework. This can be done by defining $W(\mathbf{x}|\mathbf{z}, t) = \lambda(\mathbf{z}, t)g(\mathbf{x}|\mathbf{z}, t)$ where $\lambda(\mathbf{z}, t)$ is the probability of occurrence of a jump at state $\mathbf{z}$ and $g(\mathbf{x}|\mathbf{z}, t)$ is the probability of a jump to have a specified amplitude in going from state $\mathbf{z}$ to state $\mathbf{x}$. This is exactly the process we used to produce Fig. 2.5; however, the problem is how to model $g$ and $\lambda$ based on the underlying physical phenomena [4]. Williams and Crane [10] studied the particle collision rate in turbulent flows for both solid particles and bubbles and could be a good starting point for such models. For a discussion of

multiparticle statistics in the context of kinetic theory for dense particulate flows, see [11–14].

- We deal only with Markov-type processes, which implies no dependence on past events.
- The fluid particle may be considered to be a small element of fluid with characteristic length scale smaller than the Kolmogorov length scale, $\eta = (\nu^3/\langle \varepsilon \rangle)^{1/4}$, but much larger than the molecular free path.

### 2.8.1 Definition of Distribution Function

When seeking a statistical description of a turbulent particle-laden flow, the starting point should be the nature of the phase space distribution function which describes the state of an ensemble of particles. By analogy with the kinetic theory of gases [15], it is possible to describe the ensemble by either a Klimontovich approach or a Liouville approach [16].

The Liouville approach defines a fine-grained density function conditioned on the total number of particles at a particular time $N_p(t)$. Considering the particulate phase only, where each particle is characterized by its position $\mathbf{X}^{(i)}$, velocity $\mathbf{U}^{(i)}$ and diameter $\Phi^{(i)}$, then the Liouville distribution function is defined as

$$f'(\phi_{p,1}, \mathbf{u}_{p,1}, \mathbf{x}_{p,1}, \ldots, \phi_{p,k}, \mathbf{u}_{p,k}, \mathbf{x}_{p,k}; t | N_p(t) = k)$$

$$\equiv \prod_{i=1}^{k} \delta[\phi_p - \Phi_p^{(i)}(t)] \delta[\mathbf{u}_p - \mathbf{U}_p^{(i)}(t)] \delta[\mathbf{x}_p - \mathbf{X}_p^{(i)}(t)], \qquad (2.39)$$

and the Liouville PDF is simply the ensemble of the fine-grained distribution function

$$\mathscr{P}(\phi_{p,1}, \mathbf{u}_{p,1}, \mathbf{x}_{p,1}, \ldots, \phi_{p,k}, \mathbf{u}_{p,k}, \mathbf{x}_{p,k}; t | N_p(t) = k)$$

$$= \left\langle \prod_{i=1}^{k} \delta[\phi_p - \Phi_p^{(i)}(t)] \delta[\mathbf{u}_p - \mathbf{U}_p^{(i)}(t)] \delta[\mathbf{x}_p - \mathbf{X}_p^{(i)}(t)] \right\rangle. \qquad (2.40)$$

Upon closer examination of Eqs. 2.39 and 2.40, it can be seen that the Liouville distribution is a multipoint distribution that characterizes all joint events in the ensemble. The information contained in the Liouville distribution is vast and clearly not suitable for a tractable description when deriving an engineering model. An alternative to this distribution is the Klimontovich distribution, which is defined as

$$f'(\phi_p, \mathbf{u}_p, \mathbf{x}_p; t) \equiv \sum_{i=1}^{N_p} \delta[\text{phi}_p - \Phi_p^{(i)}(t)] \delta[\mathbf{u}_p - \mathbf{U}_p^{(i)}(t)] \delta[\mathbf{x}_p - \mathbf{X}_p^{(i)}(t)], \quad (2.41)$$

and represents the number density of the particles in the phase space. The ensemble averages of the Klimontovich distribution function produce a droplet distribution function [16]

$$\mathscr{P}(\phi_p, \mathbf{u}_p, \mathbf{x}_p; t) \equiv \left\langle \sum_{i=1}^{N_p} \delta[\phi_p - \Phi_p^{(i)}(t)]\delta[\mathbf{u}_p - \mathbf{U}_p^{(i)}(t)]\delta[\mathbf{x}_p - \mathbf{X}_p^{(i)}(t)] \right\rangle,$$

(2.42)

which does not strictly represent a PDF because it does not integrate to unity over the phase space. The Klimontovich distribution function clearly contains less information (no multiparticle events) than the Liouville equation and serves as a more feasible starting point for the statistical description. In kinetic theory, there exists a simple relationship between the two descriptions (after a number of assumptions), which leads to the famous BBGKY hierarchy. However, as described by Subramaniam [16, 17], several key differences between the kinetic and particulate flows complicate the issue here. For a detailed discussion of these differences and their implications, the reader is referred to [16, 17]. For the purposes of the framework developed in this chapter, a Klimontovich distribution will be used. The distribution function will be of the form

$$\mathscr{P}(\mathbf{\Psi}_p, \mathbf{u}_p, \mathbf{x}_p, \mathbf{\Psi}_f, \mathbf{u}_f, \mathbf{x}_f; t) \equiv \left\langle \sum_{i=1}^{N_p} \delta[\psi_p - \mathbf{\Psi}_p^{(i)}(t)]\delta[\mathbf{u}_p - \mathbf{U}_p^{(i)}(t)]\delta[\mathbf{x}_p - \mathbf{X}_p^{(i)}(t)] \right.$$

$$\left. \delta[\psi_f - \mathbf{\Psi}_f^{(i)}(t)]\delta[\mathbf{u}_f - \mathbf{U}_f^{(i)}(t)]\delta[\mathbf{x}_f - \mathbf{X}_f^{(i)}(t)] \right\rangle,$$

(2.43)

where the phase space contains both the fluid and the particle. In the most general sense, the fluid and particle positions do not have to coincide ($\mathbf{X}_p^{(i)} \neq \mathbf{X}_f^{(i)}$) in Eq. 2.43 and the distribution function represents a two-point/two-particle (fluid and solid) Lagrangian distribution. However, the ultimate aim of the following analysis is to derive a single-point Eulerian model for both phases mapped onto a single spatial point via a consistency relation. Therefore, it seems more appropriate to term the following framework as a single-point, two-particle description. These descriptions are further discussed in the following sections

## 2.8.2 Eulerian and Lagrangian Two-Point, Two-Particle Description

There are two possible viewpoints of the fluid–particle system. A Lagrangian point of view describes the probability of finding two particles (fluid and discrete) at a given state $\mathbf{Z}_{fp} = (\mathbf{X}_f, \mathbf{U}_f, \mathbf{\Psi}_f, \mathbf{X}_p, \mathbf{U}_p, \mathbf{\Psi}_p)$ where $\mathbf{X}$, $\mathbf{U}$ and $\mathbf{\Psi}$ are position, velocity and an arbitrary property vectors, respectively.[3] The Lagrangian PDF is defined by $\mathscr{P}^L_{fp}(\mathbf{y}_f, \mathbf{u}_f, \boldsymbol{\psi}_f, \mathbf{y}_p, \mathbf{u}_p, \boldsymbol{\psi}_p; t)$, and the probability of finding a fluid/particle pair in the range $[\mathbf{u}_k, \mathbf{u}_k + \mathrm{d}\mathbf{u}_k]$, $[\mathbf{y}_k, \mathbf{y}_k + \mathrm{d}\mathbf{y}_k]$, $[\boldsymbol{\psi}_k, \boldsymbol{\psi}_k + \mathrm{d}\boldsymbol{\psi}_k]$ at time $t$ (where $k = f$ for fluid and $k = p$ for particle) is simply given by

$$\mathscr{P}^L_{fp}(\mathbf{y}_f, \mathbf{u}_f, \boldsymbol{\psi}_f, \mathbf{y}_p, \mathbf{u}_p, \boldsymbol{\psi}_p; t)\mathrm{d}\mathbf{y}_f\mathrm{d}\mathbf{u}_f\mathrm{d}\boldsymbol{\psi}_f\mathrm{d}\mathbf{y}_p\mathrm{d}\mathbf{u}_p\mathrm{d}\boldsymbol{\psi}_p, \qquad (2.44)$$

which conforms to the usual normalization constraint, i.e. $\int \mathscr{P}^L_{fp}\mathrm{d}\mathbf{z}_{fp} = 1$. An Eulerian (field) point of view describes the probability of finding the fluid–particle mixture in a given state $\mathbf{z}_{fp} = (\mathbf{u}_f, \boldsymbol{\psi}_f, \mathbf{u}_p, \boldsymbol{\psi}_p)$ at two fixed points $(\mathbf{x}_f, \mathbf{x}_p)$ and fixed time. Here, $\mathbf{x}$, $\mathbf{u}$ and $\boldsymbol{\psi}$ have the same meaning as in their Lagrangian counterparts, and distinction is made by using upper- and lower-case letters. Correspondingly, the probability of finding the system (at time $t$ and position $\mathbf{x}_f$, $\mathbf{x}_p$) in the given state in the range $[\mathbf{u}_k, \mathbf{u}_k + \mathrm{d}\mathbf{u}_k]$, $[\mathbf{x}_k, \mathbf{x}_k + \mathrm{d}\mathbf{x}_k]$, $[\boldsymbol{\psi}_k, \boldsymbol{\psi}_k + \mathrm{d}\boldsymbol{\psi}_k]$ is

$$\mathscr{P}^E_{fp}(\mathbf{u}_f, \boldsymbol{\psi}_f, \mathbf{u}_p, \boldsymbol{\psi}_p; \mathbf{x}_f, \mathbf{x}_p, t)\mathrm{d}\mathbf{u}_f\mathrm{d}\boldsymbol{\psi}_f\mathrm{d}\mathbf{u}_p\mathrm{d}\boldsymbol{\psi}_p. \qquad (2.45)$$

However, $\int \mathscr{P}^E_{fp}\mathrm{d}\mathbf{z}_{fp} < 1$, because in the Lagrangian point of view, positions are included in the state vector and we know that particle with the velocity $\mathbf{u}_k$ and the property $\boldsymbol{\psi}_k$ is in position $\mathbf{x}_k$ with some probability; therefore, after the integration, all particles are guaranteed to be counted. Whereas in the Eulerian point of view, we are observing the system at two fixed points $\mathbf{x}_p$ and $\mathbf{x}_f$; hence, the particle (or fluid) position is not a property of the system. Consequently, it is possible that the point $\mathbf{x}_p$, where we are expecting a discrete particle in any state $(\mathbf{u}_p, \boldsymbol{\psi}_p)$, be actually occupied by a fluid particle. Here, after the integration, some fluid or discrete particles may not be counted and the integration result will be less than unity.

### 2.8.3 One-Point Description and Consistency Relations

The above description is referred to as a two-point, two-particle description where the 'two-particle' term is added to emphasize that the two points kept in the definition of the PDF correspond to one-fluid and one-discrete-particle location. The PDF (Fokker–Planck) equation derived for this system is at least a 12-dimensional system of equations, even if only positions and velocities considered. To reduce the dimensionality, we define a one-point description with probability densities that for Eulerian description become

$$\mathscr{P}^E_f(\mathbf{u}_f, \boldsymbol{\psi}_f; \mathbf{x}_f, t) = \int \mathscr{P}^E_{fp}(\mathbf{u}_f, \boldsymbol{\psi}_f, \mathbf{u}_p, \boldsymbol{\psi}_p; \mathbf{x}_f, \mathbf{x}_p, t)\mathrm{d}\mathbf{x}_p\mathrm{d}\mathbf{u}_p\mathrm{d}\boldsymbol{\psi}_p, \quad (2.46)$$

---

[3] $\boldsymbol{\Psi}$ can be a combination of different scalars; therefore, we use a vector notation for it. However, here, we only consider the diameters, and consequently, this actually is only a scalar.

$$\mathscr{P}_p^E(\mathbf{u}_p, \psi_p; \mathbf{x}_p, t) = \int \mathscr{P}_{fp}^E(\mathbf{u}_f, \psi_f, \mathbf{u}_p, \psi_p; \mathbf{x}_f, \mathbf{x}_p, t)\mathrm{d}\mathbf{x}_f\mathrm{d}\mathbf{u}_f\mathrm{d}\psi_f. \quad (2.47)$$

This is a one-point, two-particle description, which means that we are considering the probability of finding a fluid and a discrete particle separately at two different fixed points at a fixed time, which obviously contains less information than the two-point, two-particle description where we considered the joint probability of finding a fluid and a discrete particle. This can also be mathematically justified by noting that the marginal PDFs (one-point PDFs) can always be constructed from the joint PDFs by integration, while the reverse operation is not always possible.

It should also be noted that in a conventional two-point description of a single fluid, the two different stochastic particles represent two different realizations of the flow. Thus, two fluid particles with different characteristics can exist in the same position at the same time. In the present case, the stochastic particles are two real particles (fluid and discrete) and for the Eulerian PDF, we have

$$\mathscr{P}_{fp}^E(\mathbf{u}_f, \psi_f, \mathbf{u}_p, \psi_p; \mathbf{x}_f = \mathbf{x}, \mathbf{x}_p = \mathbf{x}, t) = 0 \quad (2.48)$$

Note that a similar constraint also applies to the corresponding Lagrangian PDF. Using Eqs. (2.46) and (2.47), Eq. (2.48) results in

$$\int \mathscr{P}_f^E(\mathbf{u}_f, \psi_f; \mathbf{x}_f, t)\mathrm{d}\mathbf{u}_f\mathrm{d}\psi_f + \int \mathscr{P}_p^E(\mathbf{u}_p, \psi_p; \mathbf{x}_p, t)\mathrm{d}\mathbf{u}_p\mathrm{d}\psi_p = 1, \quad (2.49)$$

This constraint has a rather simple physical interpretation: it simply states that the whole domain is filled with the volume of all fluid particles with any velocity $\mathbf{u}_f$ and scalar property vector $\psi_f$ plus the volume of the discrete particles with any velocity $\mathbf{u}_p$ and scalar property vector $\psi_p$. Equations (2.46) and (2.47) can be normalized by defining normalization factors $\alpha_f(\mathbf{x}, t)$ and $\alpha_p(\mathbf{x}, t)$ for $\mathscr{P}_f^E$ and $\mathscr{P}_p^E$, respectively, which can be interpreted as phase volume fraction and should sum to unity.

### 2.8.4 Mass Density Function

Our ultimate goal is to derive the field (Eulerian) equations of moments of each phase by writing the Fokker–Planck equation of Eulerian distribution functions and taking the expectation (denoted by $\langle \cdot \rangle$) of a desired quantity. To do this, we need to derive the relation between the Lagrangian and Eulerian MDFs. Using Lagrangian marginal $\mathscr{P}_f^L(\mathbf{y}_f, \mathbf{u}_f, \psi_f; t)$ and $\mathscr{P}_p^L(\mathbf{y}_p, \mathbf{u}_p, \psi_p; t)$, we can define the MDF by

$$F_f^L(\mathbf{y}_f, \mathbf{u}_f, \psi_f; t) = M_f(t)\mathscr{P}_f^L(\mathbf{y}_f, \mathbf{u}_f, \psi_f; t), \quad (2.50)$$

and

$$F_p^L(\mathbf{y}_p, \mathbf{u}_p, \psi_p; t) = M_p(t)\mathscr{P}_p^L(\mathbf{y}_p, \mathbf{u}_p, \psi_p; t). \quad (2.51)$$

In these equations, $F_k^L$, with $k = f$ or $p$ for fluid and particle, respectively, can be interpreted as the probable mass of fluid or particle at the given state $\mathbf{Z_k} = (\mathbf{y}_k, \mathbf{u}_k, \psi_k)$. $M_k$ is the normalization constant for $F_k$ or the total mass of phase $k$ which physically can be calculated by $\int_{\mathcal{V}_f} \rho_f(\mathbf{x}_f, t)\mathrm{d}\mathbf{x}_f$, $\mathcal{V}_f$ being the total fluid volume, for the fluid phase, and $\sum_{i=1}^{N_p} m_{p,i}$, for the particle phase. Using these relations, we can define a two-point Lagrangian MDF as

$$\mathscr{F}_{fp}^L(\mathbf{y}_f, \mathbf{u}_f, \psi_f, \mathbf{y}_p, \mathbf{u}_p, \psi_p; t) = M_f(t)M_p(t)\mathscr{P}_{fp}^L(\mathbf{y}_f, \mathbf{u}_f, \psi_f, \mathbf{y}_p, \mathbf{u}_p, \psi_p; t)$$

$$(2.52)$$

Corresponding marginals can easily be calculated as $\mathscr{F}_f^L = M_p F_f^L$ and $\mathscr{F}_p^L = M_f F_p^L$ by integrating Eq. (2.52). At this point, it is possible to define the Eulerian fluid–particle mass density function by [4, 18]:

$$\mathscr{F}_{fp}^E(\mathbf{u}_f, \psi_f, \mathbf{u}_p, \psi_p; \mathbf{x}_f, \mathbf{x}_p, t)$$
$$= \mathscr{F}_{fp}^L(\mathbf{y}_f = \mathbf{x}_f, \mathbf{u}_f, \psi_f, \mathbf{y}_p = \mathbf{x}_p, \mathbf{u}_p, \psi_p; t)$$
$$= \int \mathscr{F}_{fp}^L(\mathbf{y}_f, \mathbf{u}_f, \psi_f, \mathbf{y}_p, \mathbf{u}_p, \psi_p; t)\delta(\mathbf{x}_f - \mathbf{y}_f)\delta(\mathbf{x}_p - \mathbf{y}_p)\mathrm{d}\mathbf{y}_f\mathrm{d}\mathbf{y}_p, \quad (2.53)$$

The marginal or one-point Eulerian MDFs are simply derived by integration of either marginal Lagrangian MDF using delta functions or just by integrating the Eulerian fluid–particle MDF over the fluid or particle phase space as follows:

$$\mathscr{F}_f^E(\mathbf{u}_f, \psi_f; \mathbf{x}_f, t) = M_p(t)F_f^E(\mathbf{u}_f, \psi_f; \mathbf{x}_f, t) \qquad (2.54)$$

and

$$\mathscr{F}_p^E(\mathbf{u}_p, \psi_p; \mathbf{x}_p, t) = M_f(t)F_p^E(\mathbf{u}_p, \psi_p; \mathbf{x}_p, t) \qquad (2.55)$$

The normalization constant for the two-point Lagrangian MDF, Eq. (2.52), is defined by $M_p M_f$. Thus, upon integrating $\mathscr{F}_{fp}^L$, we get the mass of the fluid multiplied by the mass of the discrete phase. This is done for mathematical convenience, and as a consequence, mixed indices such as $M_p F_f^L$ appear in the marginals of both two-point Lagrangian MDFs and two-point Eulerian MDFs, which are merely mathematical objects derived from Eqs. (2.50) and (2.51). However, the importance of these definitions should not be underestimated because physical quantities such as expected densities and the probabilities of presence of phases, $\alpha_f$ and $\alpha_p$, can conveniently be derived from these quantities by simple integrations:

$$\alpha_f(\mathbf{x}, t)\langle \rho_f \rangle(\mathbf{x}, t) = M_f^{-1}(t) \int \mathscr{F}_{fp}^E(\mathbf{u}_f, \psi_f, \mathbf{u}_p, \psi_p; \mathbf{x}, \mathbf{x}_p, t)\mathrm{d}\mathbf{x}_p\mathrm{d}\mathbf{u}_f\mathrm{d}\psi_f\mathrm{d}\mathbf{u}_p\mathrm{d}\psi_p$$

$$(2.56)$$

$$\alpha_p(\mathbf{x}, t)\langle\rho_p\rangle(\mathbf{x}, t) = M_f^{-1}(t) \int \mathscr{F}_{fp}^E(\mathbf{u}_f, \psi_f, \mathbf{u}_p, \psi_p; \mathbf{x}_f, \mathbf{x}, t)d\mathbf{x}_f d\mathbf{u}_f d\psi_f d\mathbf{u}_p d\psi_p$$
(2.57)

$$\alpha_f(\mathbf{x}, t) = M_p^{-1}(t) \int \rho_f^{-1} \mathscr{F}_{fp}^E(\mathbf{u}_f, \psi_f, \mathbf{u}_p, \psi_p; \mathbf{x}, \mathbf{x}_p, t)d\mathbf{x}_p d\mathbf{u}_f d\psi_f d\mathbf{u}_p d\psi_p$$
(2.58)

$$\alpha_p(\mathbf{x}, t) = M_f^{-1}(t) \int \rho_p^{-1} \mathscr{F}_{fp}^E(\mathbf{u}_f, \psi_f, \mathbf{u}_p, \psi_p; \mathbf{x}_f, \mathbf{x}, t)d\mathbf{x}_f d\mathbf{u}_f d\psi_f d\mathbf{u}_p d\psi_p$$
(2.59)

These can easily be expressed in terms of marginals of $\mathscr{F}_{fp}^E$, i.e. $F_f^E$ and $F_p^E$, by simple integration and using Eqs. (2.54) and (2.55).

# References

1. van Kampen N (2007) Stochastic processes in physics and chemistry. Elsevier, Amsterdam
2. Gardiner C (2004) Handbook of stochastic methods for physics, chemistry and natural sciences. Springer, New York
3. Oksendal B (1995) Stochastic differential equations. An introduction with applications. Springer, New York
4. Minier JP, Peirano E (2001) The PDF approach to turbulent polydispersed two-phase flows. Phys Rep 352:1–214
5. Kloeden P, Platen E (1992) Numerical solution of stochastic differential equations. Springer, New York
6. Mazur P (1959) On theory of Brownian motion. Physica 25:149–162
7. Mazur P, Oppenheim I (1970) Molecular theory of Brownian motion. Physica 50:241–258
8. Pope SB (1991) Application of the velocity-dissipation probability density function model in inhomogeneous turbulent flows. Phys Fluids A 3:1947–1957
9. Archambault MR, Edwards CF, McCormack RW (2003) Computation of spray dynamics by moment transport equations I: theory and development. Atomization Sprays 13:63–87
10. Williams JJE, Crane RI (1983) Particle collision rate in turbulent flow. Int J Multiphase Flow 9:421–435
11. Hirschfielder J, Curtiss C, Bird R (1954) Molecular theory of gases and liquids. Wiley, New York
12. Peirano E, Leckner B (1998) Fundamentals of turbulent gas-solid flows applied to circulating fluidized bed combustion. Prog Energy Combust Sci 24:259–296
13. Gidaspow (1994) Multiphase flow and fluidization. Academic Press, New York
14. Neri A, Gidaspow D (2000) Riser hydrodynamics: simulation using kinetic theory. AIChE J 46:52–67
15. Gambosi TI (1994) Gas kinetic theory. Cambridge University Press, Cambridge
16. Subramaniam S (2000) Statistical representation of a spray as a point process. Phys Fluids 12:2413–2431
17. Subramaniam S (2001) Statistical modelling of sprays using the droplet distribution function. Phys Fluids 13:624–642
18. Scott SJ (2006) A PDF based method for modelling polysized particle laden turbulent flows without size-class discretisation. PhD thesis, Imperial College, London

# Chapter 3
# Eulerian–Eulerian Field Equations

In this chapter, different methods for writing EE field equations are discussed. RANS- and PDF-type methods are introduced, and the advantages of PDF methods are discussed.

## 3.1 EE Model Development: RANS Methods

### 3.1.1 Overview

The family of RANS methods is based on a continuum view of the particle phase and was first applied by Harlow and Amsden [1]. In adopting this view, we assume that it is appropriate to treat each material as a continuum occupying the same region of space (interpenetrating). It is necessary to specify how the mixture interacts, and hence, constitutive relations for both phases and the interactions are required. The properties of a cloud of particles undergo some form of averaging process to derive Eulerian field equations for the transport of the averaged particle instantaneous properties (e.g. velocity and volume fraction) for a single realization of the flow. Several different derivations of these instantaneous equations have appeared in the literature, each using a different form of averaging. Jackson [2] employs a volume or spatial average to a finite control volume, whereas Zhang and Posperelli [3, 4] obtain similar results using an ensemble average approach. To illustrate the general method used to derive these instantaneous equations, we consider here the main features of Jackson's derivation based on the spatial average.

### 3.1.2 Instantaneous Equations

Jackson considers a suspension of spherical particles in an incompressible fluid. The relevant length scales of the system are the particle diameter $\phi_p$ and the macroscopic

J. S. Shrimpton et al., *Statistical Treatment of Turbulent Polydisperse Particle Systems*, Green Energy and Technology, DOI: 10.1007/978-1-4471-6344-2_3, © Springer-Verlag London 2014

length scale of the flow $l_f$. It is assumed that $\phi_p \ll l_f$ to ensure an acceptable separation of scales. Jackson defines a local spatial average based on a general weighting function $g(r)$, which is a function of spatial separation $r$. The integral of the weighting function over the extent of the domain is normalized to unity

$$4\pi \int\limits_0^\infty g(r)r^2 \mathrm{d}r = 1, \tag{3.1}$$

and the radius of the weighting function $a$ is defined according to the following integral relation

$$\int\limits_0^a g(r)r^2 \mathrm{d}r = \int\limits_a^\infty g(r)r^2 \mathrm{d}r, \tag{3.2}$$

where $a$ is selected to satisfy $l_f \gg a \gg \phi_p$ to ensure that averages defined by $g(r)$ do not depend significantly on the functional form of $g(r)$.

Jackson continues to define averages for the fluid phase (averaged over the fluid volume), the solid phase (averaged over the particle volumes), a mixed phase average, and a particulate phase average (summation over the particle population). Brief details of these averaging operations can be found in Appendix A. Using these averages, Jackson derives a set of instantaneous averaged equations of motion for the fluid phase, particle phase, and mixture phase. These equations contain volume averages of products, which must be decomposed in the usual way

$$\langle u_i u_j \rangle = \langle u_i \rangle \langle u_j \rangle + \langle u_i' u_j' \rangle, \tag{3.3}$$

where $u_i'$ denote the deviation of the point velocity from the phase average $u_i' = u_i - \langle u_i \rangle$. The averaged product of the velocity deviations represents a volume stress term analogous to the Reynolds stresses of single-phase turbulence. These stress terms require closure in the momentum equations for each of the fluid, solid, particulate, and mixture phases. Jackson argues that these stresses can be neglected because the particles, if sufficiently separated, do not interact via stresses induced in the fluid. Crowe et al. [5, 6] and Hinze [7] also argue that these stresses are negligible if the particle motion is governed by scales of the flow much larger than the size of the averaging volume. Following this reasoning, Jackson presents the final forms of the instantaneous averaged continuity and momentum equations for particles subject to Stokes drag.

To further simplify this discussion, we consider the case of low particle volume fraction ($\alpha_p \ll \alpha_f$) and large density ratio ($\rho_p \gg \rho_f$) and the equations are as follows

$$\frac{\partial \alpha_p}{\partial t} + \frac{\partial (\alpha_p u_{p,j})}{\partial x_j} = 0, \tag{3.4}$$

$$\frac{\partial u_{p,i}}{\partial t} + u_{p,j}\frac{\partial u_{p,i}}{\partial x_j} = \frac{1}{\tau_p}(u_{f,i} - u_{p,i}) + g_i, \tag{3.5}$$

$$\frac{\partial \alpha_f}{\partial t} + \frac{\partial \alpha_f u_{f,i}}{\partial x_j} = 0, \tag{3.6}$$

$$\frac{\partial u_{f,i}}{\partial t} + u_{f,j}\frac{\partial u_{f,i}}{\partial x_j} = -\frac{1}{\rho_f}\frac{\partial P}{\partial x_i} + \nu_f\frac{\partial^2 u_{f,i}}{\partial x_j \partial x_j} + \frac{\rho_p \alpha_p}{\rho_f \tau_p}(u_{f,i} - u_{p,i}) + g_i, \tag{3.7}$$

where $\tau_p$ represents the particle relaxation timescale for Stokes drag. It should be noted that the averaging brackets are omitted in Eqs. 3.4–3.7.

It is prudent to briefly consider the potential of the above-method for developing a transport model for poly-dispersed flows without size classes. This would be achieved by applying an appropriate averaging operation to the particle phase diameter. However, matters would be complicated by mixed diameter-velocity moments present in the momentum exchange terms requiring closure and additional volume stresses of the form $\left\langle \phi'^2_p \right\rangle$ due to deviations from the volume averaged diameter. Attempts to derive a poly-dispersed model using this approach have thus far not appeared in the literature.

### 3.1.3 Averaged Equations

The *instantaneous* averaged equations derived for the fluid and particle phases contain information regarding velocity fluctuations. To obtain averaged equations (averaged over a length scale greater than that of the instantaneous averaging of Jackson [2]) for the statistical properties of the fluctuations, a second averaging process may be applied to the instantaneous equations. The averaged equations do not contain the microscopic details of the flow, and although it is usual to apply an averaging process, it is not essential to do so (e.g. see [8]).

Averaging of the particle phase equations is analogous to the averaging of the Navier–Stokes equations in single-phase flows (e.g. Hinze [7]). The first averaging methods utilized were time and space averages, and details of subsequent refinement can be found in [9–14]. Batchelor [15] summarizes early applications of statistical averages, and Buyevich and Shchelchkova [16] provide details of applying statistical averages to the particle equations of motion.

Numerous different averaging types are discussed in the literature [17], and a brief summary of the main methods are discussed here. Firstly, let $\psi_k(\mathbf{x}, t)$ represent an instantaneous microscopic field in phase $k$ and $\langle \psi_k(\mathbf{x}, t) \rangle$ represents the average value of this field following an arbitrary averaging process. Using this notation, a simple time average is defined as

$$\overline{\psi}(\mathbf{x}, t) = \langle \psi(\mathbf{x}, t) \rangle_T = \frac{1}{T} \int_T \psi(\mathbf{x}, t) dt, \qquad (3.8)$$

where the field is averaged over a time interval $T$. In a similar fashion, a volume average is given as the integral of the field over a volume $V$

$$\langle \psi(\mathbf{x}, t) \rangle_V = \frac{1}{V} \int_V \psi(\mathbf{x}, t) d\mathbf{x}. \qquad (3.9)$$

The ensemble (statistical) average used by Shih and Lumley [18] is formed in a different manner, where the field is averaged over $N$ realizations

$$\langle \psi(\mathbf{x}, t) \rangle_E = \lim_{N \to \infty} \frac{1}{N} \sum_{i=1}^{N} \psi^{(i)}(\mathbf{x}, t). \qquad (3.10)$$

Also found in the literature [17] are phasic and mass weighted averages. Drew [17] defines a phase indicator function[1] $\chi_k(\mathbf{x}, t)$ where

$$\chi_k(\mathbf{x}, t) = \begin{cases} 1 \text{ if } \mathbf{x} \text{ is in phase } k \text{ at time } t, \\ 0 \text{ otherwise.} \end{cases} \qquad (3.11)$$

The phasic average is defined as

$$\langle \psi(\mathbf{x}, t) \rangle_{V_k} = \frac{1}{V_k} \int_{V_k} \psi(\mathbf{x}, t) d\mathbf{x} = \frac{\langle \chi_k \psi \rangle_V}{\langle \chi_k \rangle_V}, \qquad (3.12)$$

and similarly, the mass weighted average is given by

$$\langle \psi(\mathbf{x}, t) \rangle_{\rho_k} = \frac{1}{\rho_k V_k} \int_{V_k} \psi(\mathbf{x}, t) d\mathbf{x}. \qquad (3.13)$$

Following the averaging process, all variables are decomposed into mean values $\overline{\psi}$ and the fluctuation/deviation from this mean value $\psi'$ in the spirit of the single-phase Reynolds average

$$\psi = \overline{\psi} + \psi'. \qquad (3.14)$$

This decomposition results in the final form of the averaged two-fluid equations for both phases [19]

$$\frac{\partial \overline{u}_{f,j}}{\partial x_j} = 0, \qquad (3.15)$$

---

[1] The author believes that the phase indicator function is not strictly applicable for averaging the instantaneous equations that already represent a continuum.

$$\frac{\partial \overline{u}_{f,i}}{\partial t} + \frac{\partial \overline{u}_{f,j} \overline{u}_{f,i}}{\partial x_j} = -\frac{1}{\rho_f} \frac{\partial P}{\partial x_i} + \nu_f \frac{\partial^2 \overline{u}_{f,i}}{\partial x_j \partial x_j} - \frac{\partial \overline{u'_{f,i} u'_{f,j}}}{\partial x_j}$$

$$- \frac{\rho_p \overline{\alpha}_p}{\rho_f \tau_p} (\overline{u}_{f,i} - \overline{u}_{p,i}) - \frac{\rho_p}{\rho_f \tau_p} (\overline{u'_{f,i} \alpha'_p} - \overline{u'_{f,i} \alpha'_p}) + g_i,$$

(3.16)

$$\frac{\partial \overline{\alpha}_p}{\partial t} + \frac{\partial (\overline{u}_{p,j} \overline{\alpha}_p)}{\partial x_j} = -\frac{\partial \overline{u'_{p,j} \alpha'_p}}{\partial x_j},$$

(3.17)

$$\frac{\partial \overline{u}_{p,i}}{\partial t} + \overline{u}_{p,j} \frac{\overline{u}_{p,i}}{\partial x_j} = -\frac{\partial \overline{u'_{p,i} u'_{p,j}}}{\partial x_j} + \frac{1}{\tau_p} (\overline{u}_{f,i} - \overline{u}_{p,i}) + \overline{u'_{p,i} \frac{\partial u'_{p,j}}{\partial x_j}} + g_i.$$

(3.18)

The above averaged equations highlight the seminal turbulence closure problem for the Reynolds stresses and the volume fraction fluxes. In this form, these stresses require modeling of some form. However, the volume fraction fluxes can be eliminated if a weighted average is used [19] analogous to Favre averaging [20] for single-phase turbulence. In the Favre averaging process, the velocities are weighted by the fluid density. Here, it is possible to use the volume fraction as the weighting factor. Denoting the "Favre" averaged quantity as $\tilde{\psi}$, it is related to the instantaneous value by the relation

$$\psi = \tilde{\psi} + \psi'',$$

(3.19)

where $\psi''$ is the fluctuation about the Favre averaged value. The Favre average is calculated according to

$$\tilde{\psi} = \frac{1}{\alpha} \overline{\alpha \psi} = \frac{1}{\alpha T} \int_T \alpha \psi \, dt.$$

(3.20)

Using this approach, the particle continuity equation becomes

$$\frac{\partial \alpha_p}{\partial t} + \frac{\partial (\alpha_p \tilde{u}_{p,j})}{\partial x_j} = 0,$$

(3.21)

and the closure problem no longer features in the transport equation. However, to recover the Reynolds averaged value of a quantity, the unclosed flux is again required

$$\overline{\psi} = \frac{1}{\alpha} (\alpha \tilde{\psi} - \overline{\alpha' \psi'}).$$

(3.22)

It is this closure problem that has motivated the majority of research in the RANS field for two-phase flow. In general, the methods published in the literature parallel the methods developed for single-phase flow, with extensions to close the additional terms due to the particle phase. Given the extensive effort that has been devoted to the single-phase techniques, this would seem the sensible path to take. The next sections discuss briefly the various approaches that have been developed, namely eddy viscosity methods, second moment closure methods and algebraic closures.

## 3.1.4 Eddy Viscosity Models

The simplest two-fluid models are based on the single-phase turbulent viscosity models such as the $k - \varepsilon$ model [21]. In the turbulent viscosity approach for single-phase flow, it is conventional to model the anisotropic Reynolds stresses using the mean velocity gradient and a turbulent viscosity. One early example of a two-fluid $k - \varepsilon$ model for particulate flows is that of Pourahmadi and Humphrey [22]. Pourahmadi and Humphrey [22] extend the single-phase approach to model the Reynolds stresses in the dispersed phase and the turbulent volume fraction fluxes. The dispersed phase turbulent kinetic energy and the trace of the fluid–particle co-variance tensor were related to the carrier phase kinetic energy according to

$$k_p = k_f \frac{T_L}{\tau_p + T_L} \quad \text{and} \quad \overline{u'_{f,i} u'_{p,i}} = 2k_f \frac{T_L}{\tau_p + T_L}, \tag{3.23}$$

where $T_L$ is the fluid Lagrangian integral timescale. According to Pourahmadi and Humphrey [22], these expressions are valid for high Reynolds numbers and small particle time constants. The model proposed only accounts for turbulent fluctuations in the dispersed phase induced by the fluid phase. Contributions to the particle phase turbulence due to a mean velocity gradient in the dispersed phase are neglected.

Elghobashi and Abou-Arab [23] present a two-equation model for predicting two-phase flows. The two equations proposed model the transport of kinetic energy and rate of dissipation in the carrier phase of the two-phase flow, and both are derived directly from the momentum equation of the carrier fluid according to a volume averaged approach. Neglecting particle collisions, but accounting for two-way coupling, Elghobashi and Abou-Arab [23] identify the unclosed correlations following a Reynolds decomposition and time averaging. The unclosed terms account for Reynolds stresses, volume fraction fluxes, pressure interaction, and strain rate correlations. These unclosed terms are modeled according to eddy viscosity approaches, and the volume fraction flux is modeled in a manner analogous to methods used for passive scalars. In Elghobashi et al. [24], test the two-equation model against experimental data for a particle laden turbulent axisymmetric jet flow. The model shows good agreement with the experimental data for the mean flow properties, the turbulence kinetic energy and carrier flow shear stresses. The next stage in the model development involved evaporating droplets in isothermal flow [25]. The carrier flow was assumed incompressible, and droplet evaporation was due only to the vapor concentration gradient. The distribution of particle sizes was accounted for by using separate size classes, each modeled as a separate interpenetrating phase. This method of accounting for particle size distributions (introduced by Tambour [26]) is discussed in more detail later in this chapter. The results from this investigation again showed good agreement with available experimental data. Rizk and Elghobashi [27] further extend the model to low Reynolds number flows with wall boundary conditions. The model is tested against experimental data for a turbulent particle laden pipe flow, and good agreement is shown.

The Chen and Wood model [28] employ a separate eddy viscosity for the continuous and dispersed phases. The dispersed phase eddy viscosity was modelled according to the Hinze–Tchen relation

$$v_p^t = \frac{v_f^t}{1 + (\tau_p/T_\varepsilon)}, \tag{3.24}$$

where $T_\varepsilon$ is a turbulence timescale and $v_f^t$ and $v_p^t$ are the fluid and particle turbulent viscosities, respectively, all of which require modeling. Mashayek and Taulbee [29] assess the accuracy of the Hinze–Tchen relation with DNS of homogenous shear flow and experimental data for an axisymmetric jet. The Hinze–Tchen relation is shown to deviate significantly from the DNS and experimental data. Mashayek and Taulbee [30] explain this deviation in terms of the fundamental model assumption that the particle stays within the same fluid element for the whole of the particle motion. This assumption is only valid for small particle relaxation times.

The two-equation model was also implemented by Issa and Olivera [31] with the addition of dispersed phase transport equations. Again, the turbulent stresses of each phase were modeled using the eddy viscosity assumption, but in this example, the dispersed phase stresses were assumed proportional to the carrier phase stresses.

Picart et al. [32] use a set of implicit algebraic (for the carrier phase) relations to form a $k - \varepsilon$ model. The dispersed phase is modeled using a dispersion tensor derived in the framework of Tchen's theory [33, 34].

It is worth noting that not all $k - \varepsilon$ models are restricted to two-equation models. For example, Abou-Arab and Roco [35] derive a one-equation model for the continuous phase turbulence kinetic energy, which accounts for turbulence modification by the particles. Conversely, Mashayek, and Taulbee [30] derive a four-equation model based on an explicit algebraic stress model.

Lain and Aliod [36] derive a $k - \varepsilon$ model based on the dispersed elements PDF indicator function ensemble conditioned average of Aliod and Dopazo [37]. A standard $k - \varepsilon$ approach is used for the carrier phase, and a kinetic energy equation is also derived for the particulate phase. The turbulent fluxes in the particulate phase are modeled according to an anisotropic Boussinesq–Prandtl-type closure which accounts for enhanced particle stresses in the stream-wise direction, which is consistent with the theoretical analysis of Reeks [38]. The model of Lain and Aliod showed good agreement with experimental data for a particle laden jet flow.

Zhou et al. [39, 40] derive a non-linear $k - \varepsilon - k_p$ model with an anisotropic particle eddy viscosity tensor for the dispersed phase. Zhou remarks that $k - \varepsilon$ models have been relatively successfully employed in the simulation of non-swirling and weakly swirling flows, but for more complex flows, second moment models are more successful. Zhou et al. compare results from the non-linear model with experimental data for a swirling gas particle flow. Zhou concludes that the non-linear model has modeling capability comparable with the second-order models, with significantly less computation demand.

Despite the apparent success and convenience of the eddy viscosity type models for particulate flows, evidence suggests that the dispersed phase is in fact highly anisotropic [41–43] and the an isotropic approach is not well justified. This has encouraged development of more sophisticated models such as second moment closures which are now discussed.

## 3.1.5 Second Moment Closure Models

Second moment closure models are significantly more complex than their turbulent viscosity model counterparts. For this reason, there has been less research published in this area. Nevertheless, second moment models generally provide more accurate predictions of complex flows due to the anisotropic nature of the dispersed phase [42]. Second moment closure models require transport equations for the statistical moments up to second order, in addition to closures for the higher order correlations. The range of second moment closure models for single-phase flow is extensive [44] and will not be discussed here. Instead, a brief summary of some of the more significant two-phase turbulence models will be given.

Shih and Lumley [18] present a second-order model based on the assumption that the particle size and time constant were smaller than the carrier phase turbulent length and timescales, respectively. The convective term in the particle equation was neglected, and the particle velocity was formed from an expansion of the carrier phase velocity. This assumption leads to equivalence between the particle and carrier phase Reynolds stresses, which is questionable when considering particle dispersion mechanisms such as crossing trajectories. The transport equation for the particle density fluctuations was augmented by a dissipative term to assure bounded solutions for the numerical method. Shih and Lumley [18] test their model using experimental data for a particle mixing layer in decaying homogeneous turbulence for both heavy and light particles. The results show that the crossing trajectory effect is a more dominant mechanism in particle dispersion than inertia effects.

Simonin et al. [45, 46] propose second moment models that include interparticle collisions and two-way coupling. Modification of the carrier phase turbulence field was accounted for by inclusion of source terms in the transport equations for the fluid turbulence kinetic energy and dissipation rate. An equation was derived for the velocity covariance of the fluid and particle phase velocities. This equation was modeled using an anisotropic eddy viscosity method which allowed anisotropy and anti-symmetry in the covariance tensor. Particle collisions were accounted for in the particle Reynolds stress equation using an additional source term which was derived from Grad's theory [47]. The models also accounted for turbulent transport of particle Reynolds stresses, production via mean velocity gradients in the particle phase and inertia–drag effects. The third-order correlations were closed by deriving a transport equation for the moments and neglecting the convection terms.

Lopez de Bertodano et al. [48] derive an intermediate model based on the instantaneous equation as derived by Ishii [13] for bubbly flows. A Reynolds stress model for the carrier phase accounting for two-way interactions is derived, but the dispersed phase is modeled according to an isotropic approximation. This apparent discrepancy in information between the two phases is justified on the grounds that the bubbly phase momentum equation is dominated by the pressure gradient and interfacial forces. Hence, the dispersed phase Reynolds stresses are small and can be neglected.

Lain and Aliod [49] build upon their previous $k - \varepsilon$ model to derive a full two-fluid Reynolds stress model, again based on the instantaneous equations as derived by Aliod and Dopazo [37]. The model considers non-colliding, high inertial particles in an axisymmetric flow. The carrier phase is modeled using an algebraic Reynolds stress model (ARSM), which is further discussed in the next section. The ARSM used was that of Picart et al. [32], which required an initial $k - \varepsilon$ solution for the carrier flow, then an iterative procedure to calculate the Reynolds stresses. Lain and Aliod [49] use their previous $k - \varepsilon$ model [36] in conjunction with the iterative procedure to calculate the carrier flow. Two-way coupling is accounted for using interaction source terms based on an empirical particle drag law valid for particle Reynolds numbers up to 1,000. The two-phase velocity covariance tensor was closed using a functional expression rather than a transport equation. The particle shear stresses are modeled in the limit of high-inertia particles, and it is assumed that the particle diffusivity tensor will be reduced to a scalar quantity related to the normal stresses. This simplifies the set of particle Reynolds stress equations to just the normal stress components.

## 3.1.6 Algebraic Models

The Reynolds stress models offer significant improvements in prediction accuracy over the more simple eddy viscosity type model. However, this improvement is accompanied in a large increase in complexity in both mathematical description and computational implementation. In the single-phase community, an intermediate class of model has been developed, known as the algebraic Reynolds stress model (ARSM). In this class of model, closed algebraic (either explicit or implicit) expressions for the Reynolds stresses are derived and the model is significantly simpler to implement for engineering-type applications. Considering initially single-phase ARSM, one of the initial models was that developed by Rodi [50]. This model was an implicit model, and so required an iterative solution method. Pope [51] introduced an explicit model that negated the need for costly iterative solutions, but was only applicable to two-dimensional flows. The extension to three dimensions was completed by Taulbee [52] and Gatski and Speziale [53]. ARSM for two-phase flows are more recent development, and the number of models remains few. Mashayek and Taulbee [54] extend the single-phase analysis to include the particulate phase. The details of the derivation are beyond the scope of this review, but Mashayek [19] summarizes the main steps as (1) construction of differential transport equations for the second-order moments; (2) modeling of the unclosed terms in the transport equations (pressure-strain, pressure-void fraction gradient correlations); (3) simplification of the transport equations to yield implicit algebraic relations; and (4) solution of the implicit algebraic equations to generate explicit models.

## 3.2 EE Model Development: PDF Methods

### 3.2.1 Overview

The probability density function (PDF) approach was pioneered by the work of Buyevich [55–57] and Reeks [38, 58–60]. The method is analogous to the kinetic theory developed for gases [61]. The start point for the PDF method is the particle phase space density function defined in the particle phase space. The particle is defined in terms of a Lagrangian state vector $Z_i$ and phase space vector $z_i$[2]

$$Z_i = (\mathbf{X}_p, \mathbf{U}_p, \mathbf{\Psi}_p) \quad \text{and} \quad z_i = (\mathbf{x}_p, \mathbf{u}_p, \psi_p), \quad (3.25)$$

which include particle phase space properties such as position $x_{p,i}$ and velocity $u_{p,i}$. It should be noted that fluid properties are not included in the particle state vector. The corresponding phase space density function takes the form of a Klimontovich distribution [62], and is defined [19] as

$$W(z_i, t) = \prod_{i=1}^{N} \delta[Z_i(t) - z_i], \quad (3.26)$$

where $\delta$ is the Dirac delta function. The particle trajectory through the phase space is governed by the Lagrangian equation of motion [19]

$$\frac{dZ_i}{dt} = K_i(Z_1, Z_2, \ldots, Z_N, t). \quad (3.27)$$

The PDF for the particle in phase space is the ensemble average of the density function (3.26) over all phase space dimensions

$$\mathscr{P} = \langle W(z_i, t) \rangle = \left\langle \prod_{i=1}^{N} \delta[Z_i(t) - z_i] \right\rangle. \quad (3.28)$$

---

[2] The general notation adopted from this point onward is that Eulerian variables are denoted by lower-case characters, and Lagrangian variables are denoted by upper-case characters.*****

Differentiation of Eq. 3.26 with respect to time [63] or use of the Liouville equation [64] results in

$$\frac{\partial W(z_i, t)}{\partial t} = \sum_{i=1}^{N} \frac{dZ_i(t)}{dt} \frac{\partial}{\partial z_i} \prod_{j=1}^{N} \delta[Z_j(t) - z_j], \qquad (3.29)$$

from which, following a number of simplifications, the final form is obtained as follows

$$\frac{\partial W}{\partial t} + \frac{\partial}{\partial z_i}[K_i(z_1, z_2, \ldots, z_N)W] = 0, \qquad (3.30)$$

which indicates that the phase space density is conserved (in the absence of particle collisions) along the particle trajectory in phase space. Taking the ensemble average of Eq. 3.30 results in

$$\frac{\partial \langle W \rangle}{\partial t} + \frac{\partial \langle K_i \rangle \langle W \rangle}{\partial z_i} + \frac{\partial \langle K_i' W \rangle}{\partial z_i} = 0, \qquad (3.31)$$

which demonstrates the closure problem due to the correlation between the fluctuation $K_i'$ with the phase space density function $W$. Significant research effort has been expended in search of closure for this correlation. Most of the proposed closures are developments from theories developed for single-phase turbulence such as Kraichnan's direct interaction approximation (DIA) [65] and Lagrangian history direct interaction approximation (LHDIA) [66]. The main approaches within the two-phase literature are now discussed in further detail.

### 3.2.2  DIA and LHDIA Closure

The application of DIA and LHDIA to the kinetic equation for particulate flows was pioneered by Reeks. Considering a particle subject to a Stokes drag force, Reeks derived [58] the following form of the kinetic equation

$$\frac{\partial \langle W \rangle}{\partial t} + \frac{\partial}{\partial x_i} \left[ u_{p,i} \langle W \rangle \right] + \frac{\partial}{\partial u_{p,i}} \left[ \frac{1}{\tau_p} \left( \langle u_{f,i} \rangle - u_{p,i} \right) \langle W \rangle \right] = -\frac{\partial}{\partial u_{p,i}} \left[ \frac{1}{\tau_p} \langle u_{f,i}' W \rangle \right]. \qquad (3.32)$$

The last term on the right-hand side of Eq. 3.32 represents the phase space "diffusion" current. Reeks recognizes that Eq. 3.32 is linear in $W$, so a solution can be obtained via a general Green's function $G$ and initial condition for the phase space density function $W(\mathbf{x}^0, \mathbf{u}_p^0, t^0)$. Given these assumptions, the phase space density at any point in time is given by

$$W(\mathbf{x}, \mathbf{u}_p, t) = \int G(\mathbf{x}, \mathbf{u}_p, t; \mathbf{x}^0, \mathbf{u}_p^0, t^0) W(\mathbf{x}^0, \mathbf{u}_p^0, t^0) d\mathbf{x}^0 d\mathbf{u}_p^0. \tag{3.33}$$

If it is further assumed that $G$ and $W$ are not correlated, then it is trivial to take the ensemble average of Eq. 3.33 and obtain

$$\mathscr{P} = \langle W(\mathbf{x}, \mathbf{u}_p, t) \rangle = \int \langle G(\mathbf{x}, \mathbf{u}_p, t; \mathbf{x}^0, \mathbf{u}_p^0, t^0) \rangle \langle W(\mathbf{x}^0, \mathbf{u}_p^0, t^0) \rangle d\mathbf{x}^0 d\mathbf{u}_p^0, \tag{3.34}$$

where the ensemble average of the product $\langle GW \rangle$ is separable due to the assumption of no correlation between the functions. This result implies that the solution to $\langle W \rangle$ is determined by the average Green's function. Continuing this analysis results in

$$\frac{\partial \langle G \rangle}{\partial t} + \frac{\partial}{\partial x_i} \left[ u_{p,i} \langle G \rangle \right] + \frac{\partial}{\partial u_{p,i}} \left[ \frac{1}{\tau_p} \left( \langle u_{f,i} \rangle - u_{p,i} \right) \langle G \rangle \right] = -\frac{\partial}{\partial u_{p,i}} \left[ \frac{1}{\tau_p} \langle u'_{f,i} G \rangle \right], \tag{3.35}$$

in which the closure problem has now become $\langle u'_{f,i} G \rangle$. Reeks closed this expression using Kraichnan's DIA method [65], which provides closure for the phase space diffusion current.

Unfortunately, Reeks showed [58] that a number of inconsistencies exist in the DIA solution he proposed when considering a situation with zero mean fluid velocity and constant fluid fluctuating velocity (but random for each realization). This was shown to be a consequence of the solution's incompatibility with a random Galilean transform (RGT), and this led Reeks to derive [59] a modified general form for the phase space diffusion current

$$\frac{1}{\tau_p} \langle u'_{f,i} W \rangle = - \left( \lambda_{ji} \frac{\partial}{\partial x_j} - \mu_{ji} \frac{\partial}{\partial u_{p,j}} \right) \langle W \rangle. \tag{3.36}$$

In Eq. 3.36, $\lambda_{ij}$ and $\mu_{ij}$ are phase space diffusion tensors and the reader is referred to [59] for their complete forms.

The general expression for the phase space diffusion current is again derived by Reeks, but this time using LHDIA [60]. An extended expression (compatible with RGT) for inhomogeneous turbulent flows was obtained as

$$\frac{1}{\tau_p} \langle u'_{f,i} W \rangle = - \left( \lambda_{ji} \frac{\partial}{\partial x_j} - \mu_{ji} \frac{\partial}{\partial u_{p,j}} + \gamma_i \right) \langle W \rangle, \tag{3.37}$$

which is similar to the DIA expression with the addition of a term $\gamma_i$ that accounts for the drift of the phase space PDF due to inhomogeneities in the fluid phase turbulence. Although the closed form of the PDF equation is now realized, specific analytical solutions for the phase space diffusion and drift tensors are still required. Hyland et al. [67] derived an exact analytical solution for particles dispersed in a homogenous turbulent shear flow.

### 3.2.3 Furutsu–Novikov–Donsker Closure

Closed forms of the PDF kinetic equation can also be obtained using a method based on functional calculus. Several authors [63, 68–71] use the Furutsu–Novikov–Donsker formula (FND) to obtain similar results to those of Reeks [60]. Derived independently by Furutsu, Novikov and Donsker, the FND formula is applicable to any random Gaussian function $g_i(p)$ and shows that the correlation of $g_i(p)$ with a function $F[g_i]$ which is itself a function of $g_i$ can be given in terms of a functional derivative [19]

$$\langle g_i(p)F[\mathbf{g}] \rangle = \int \langle g_i(p)g_j(p') \rangle \left\langle \frac{\delta F[\mathbf{g}]}{\delta g_j(p')\mathrm{d}p'} \right\rangle \mathrm{d}p'. \tag{3.38}$$

The assumption that the fluid velocity fluctuations conform to a Gaussian distribution is questionable [72]. However, as pointed out by Mashayek and Pandya [19], the FND derivation serves as a valid first test for other closure schemes in the limiting case of Gaussian velocity fluctuations. This functional approach is the basis of work from Derevich [68, 69], Zaichik et al. [70, 73–77], Hyland et al. [63] and Pandya and Mashayek [71]. Hyland et al. [63] uses the FND formula to obtain a closure identical to the LHDIA closure of Reeks.

In [63], the transport equation for the particle phase space density is derived using a Langevin-type equation, but using a non-white noise fluctuating driving force. This results in a form of the Liouville equation identical to that found in [59] and includes the unclosed phase space diffusion current. Hyland et al. go on to derive the closed form of the diffusion current for a Gaussian driving force using the FND formula.

Zaichik [74] derived a kinetic PDF equation for turbulent particulate flows with heat transfer. This results in an additional unclosed phase space diffusion current due to the correlation between the particle fluctuating temperature and the particle phase space density function. Zaichik uses the FND formula (again assuming a Gaussian fluctuating driving force) to propose a closure to the new term and goes on to derive transport equations for moments of the particle properties, up to and including the third-order moments. In order to find a closed set of equations, Zaichik proposes algebraic relations for the third-order correlations by neglecting the time evolution, convection and generation due to mean velocity and temperature gradients.

Derevich [68] again uses the FND formula to close the PDF equation and obtain a set of moment equations. Wall boundary conditions are derived using an approximate solution to the PDF equation, which describe a loss of particle impulse during collisions with the wall. In [69], Derevich tests the model for a vertical particle laden pipe flow and demonstrates that the model is capable of capturing several physical phenomenon including (1) increase of particulate kinetic energy above that of the fluid near to the wall; (2) the particle deposition velocity at the wall is controlled by turbophoretic forces; and (3) the intensity of the particulate fluctuations in the axial direction exceeds that of the radial directions.

### 3.2.4 Van Kampen Closure

The Van Kampen (VK) method [78, 79] proposes a cumulant expansion method for solution of a set of linear stochastic differential equations of the form

$$\frac{\mathrm{d}\mathbf{Z}(t)}{\mathrm{d}t} = [A_0 + \alpha \langle A_1(t) \rangle] \mathbf{Z}(t), \tag{3.39}$$

where $A_0$ and $A_1$ are linear operators and $\alpha$ is the magnitude of the stochastic fluctuations.

Application of Van Kampen's method in the field of multi-phase flow was first proposed by Pozorski and Minier [80] to close the PDF kinetic equation for simple homogenous turbulence and the more complicated case of non-homogenous turbulence. Pozorski and Minier present results identical to those derived by Reeks [60] using LHDIA.

Pandya and Mashayek [71] use Van Kampen's method to obtain a closed form of the PDF equation for collision-free evaporating droplets in an isothermal isotropic flow. The resulting Fokker–Planck-type equation is solved using the well-established path-integral method in particle phase space. It should be noted that the macroscopic equations for the statistical moments of the particulate properties are not derived in this work.

In a different work, Pandya and Mashayek [81] consider evaporating droplets in non-isothermal flows. A closed form of the PDF equation is derived using both LHDIA and Van Kampen's method, and the resulting closure is shown to be consistent with an extended form of RGT. Eulerian transport equations are derived from the PDF equation, and the solutions of these equations show good correlation with DNS results for simple shear flows (velocity and temperature).

### 3.2.5 Langevin Models

In this section, dispersed phase closures based on the Langevin equation are discussed. Specifically, it is closures using a white noise Wiener process that is of principal interest here. The use of the Wiener process dictates a number of assumptions regarding the nature of the turbulent fluctuations, which are not assumed in other PDF models. For this reason, it can be argued that Wiener-based PDF methods are not as mathematically rigourous and physically justified as alternative methods. Despite this drawback, this class of method provides a powerful and flexible approach toward the problem of modeling two-phase particulate flows.

The Wiener-based Langevin equation was originally proposed (in the context of particulate flows) by Simonin et al. [82, 83]. However, much of the initial work developing a framework was published by Simonin [84] a number of years earlier. In [84], Simonin considers a dilute suspension of mono-dispersed particles with collisions. Starting initially from a set of instantaneous field equations, a density-weighted averaged set of equations for the particulate phase was derived. Granular stresses due to particle collisions are assumed negligible in comparison with the particle Reynolds stresses, and the dispersion effects due to correlations between carrier phase pressure and particle distribution are assumed negligible compared to fluctuations of the particle drag force. Following these assumptions, modeling of the interfacial momentum transfer becomes of primary importance. Simonin formulates the drag force in terms of a relative velocity

$$v_{r,i} = \langle u_{p,i} - u_{f,i} \rangle_p$$
$$\langle u_{p,i} \rangle_p - \langle u_{f,i} \rangle_p$$
$$\langle u_{p,i} \rangle_p - \langle u_{f,i} \rangle_f - v_{d,i}, \tag{3.40}$$

where $\langle u_f \rangle_p$ represents the fluid velocity averaged over each point in the particulate phase. The drift velocity $v_d$ represents a turbulent dispersion effect due to the transport of particles due to fluctuating fluid motion. This term is modelled as

$$v_{d,i} = -D^t_{fp,ij} \left[ \frac{1}{\alpha_p} \frac{\partial \alpha_p}{\partial x_j} - \frac{1}{\alpha_f} \frac{\partial \alpha_f}{\partial x_j} \right], \tag{3.41}$$

where $D^t_{fp,ij}$ is a turbulent fluid–particle correlation tensor written as

$$D^t_{fp,ij} = \tau^t_{fp} \langle u'_{f,i} u'_{p,j} \rangle_p. \tag{3.42}$$

The eddy particle interaction timescale $\tau^t_{fp}$ is modeled in an approach consistent with Csanady [85] and accounts for the crossing trajectory effects [86]. The fluid–particle covariance tensor is modeled using a simple eddy viscosity approach.

One of the key considerations of the approach described above is the question of how to model the fluid velocity sampled by the particle along its trajectory. In [82, 83], Simonin et al. address this problem by means of a Langevin equation which is a modification of the General Langevin Model (GLM) of Haworth and Pope [87]. The Simonin model includes the additional correction term $(U_{p,j} - U_{f,j}) \frac{\langle u_{f,i} \rangle}{\partial x_j} dt$ which accounts for the relative displacement between the fluid element and the particle position. This allows the dispersion characteristics to be closed at the more fundamental Lagrangian level, rather than at the Eulerian macroscopic level. Simonin et al. [45] compare predictions from this model with results from LES for simple shear flows. It is found that the inertia effect leads to an increase in anisotropy of the particle fluctuating motion with respect to the fluid motion, and this was reproduced by the Langevin based model.

In [88, 89], Simonin obtains a set of two-fluid equations via a kinetic-type equation for the particulate phase space, rather than the instantaneous field equations of earlier works. The framework is extended to include non-isothermal flows with evaporating particles/drops.

Pozorski and Minier [80] demonstrate the relationship between a formal kinetic approach akin to Reeks [38, 59, 60] and that of a Langevin approach with a white noise Wiener process with diffusion tensor $B_{ij}$. It is shown that if the dispersion tensors $\lambda_{ij}$ and $\mu_{ij}$ are assumed constant (for simplicity), then the particle Langevin diffusion tensor $B_{ij}$ can be expressed as

$$\mathbf{B} = \begin{pmatrix} 0 & \lambda_{ij} \\ \lambda_{ji} & \mu_{ij} + \mu_{ji} \end{pmatrix}. \tag{3.43}$$

Pozorski and Minier [80] go on to explain that the difference between the formal kinetic and Langevin approaches is due to the fact that $B_{ij}$ is not positive definite, which shows that the formal kinetic approach is not Markovian, whereas the Wiener process is (i.e. no history forces). Despite this inconsistency, closure of the formal kinetic approach is problematic because the fluid velocity is an external variable. This motivates inclusion of the fluid velocity, and the fluid velocity sampled by the particle, into the system state vector for the phase space. Minier and Pozorski [80] remark that despite the previous development for Langevin models of this nature, they have not been mathematically rigourous or physically justified. Using a Wiener process is only justified for fast fluctuating variables when the characteristic timescale is much smaller than the scale of the system. Application of Kolmogorov theory suggests that for high Reynolds number flows, the fluid accelerations are uncorrelated (in relation to the inertia scales), whereas the velocity correlation remains correlated [90]. While this goes some way toward justifying a Wiener process for the fluid Langevin equation, the nature of the Langevin equation describing the velocity sampled (or "seen") by the particles remains open for debate [91].

Peirano and Minier [92, 93] extended the framework introduced in [80] to demonstrate a hierarchy between the classical Lagrangian and Eulerian approaches within the general PDF approach. A two-point description that includes both the fluid and particulate phases is formalized in terms of Lagrangian and Eulerian fields. Macroscopic transport equations for various statistical moments are derived, and several simplifications are proposed.

Elperin et al. [94–96] use a Wiener process to describe the fluctuating components of the particle trajectories. This restricts the findings to particles with rapid uncorrelated fluctuations in position. Using this approach, it is demonstrated that in a turbulent flow with a mean temperature gradient, there exists a flux of particles in the opposite direction to that of the temperature gradient.

## 3.3 Poly-dispersed Methods

The literature available for polydispersed methods is far less than is available for monodispersed particulate flows. In general, polydispersed methods can be categorized as either sectional or non-sectional methods. In this section, both approaches are discussed and the advantages and disadvantages of both are highlighted. Where possible, links are made between the following polydispersed approaches and the previously discussed monodispersed techniques.

### 3.3.1 Sectional Models

The first sectional approach was first proposed by Tambour [26]. In this approach, the particle population is sectioned into a series of 'bins', each corresponding to a particular diameter range $[\phi_p^{(i)}, \phi_p^{(i)} + \delta\phi_p]$. Each size class is treated as a separate but interpenetrating phase with a sectional conservation equation for mass and momentum. Tambour focusses on the inter and intrasectional source terms for vaporization and coalescence.

The work of Tambour was later formalized in relation to the spray equation of Williams [97, 98] by Greenberg et al. [99]. The spray equation is analogous to the Boltzman equation for gasses [61] and can be considered a kinetic-type equation for a particle population. Williams considered a general ensemble of particles characterized by position, velocity, radius, temperature, and time, and derived a conservation equation for the particle distribution function in phase space

$$\frac{\partial F}{\partial t} + \frac{\partial}{\partial x_k}[u_k F] + \frac{\partial}{\partial u_k}\left[\langle A_k | \mathbf{x}, \mathbf{u}, \phi_p; t\rangle F\right] + \frac{\partial}{\partial \phi_p}\left[\langle \Theta | \mathbf{x}, \mathbf{u}, \phi_p; t\rangle F\right] = 0, \quad (3.44)$$

where $\mathbf{A}$ is a drag source term and $\Theta$ is an evaporation source term[3]. It should be noted that the Williams equation does not contain information regarding the carrier phase (similar to the PDF methods of Sect. 3.2), which is necessary for calculation of $\mathbf{A}$ and $\Theta$. In general, the carrier phase contributions to the drag and evaporation source terms must be determined using a model of some form. Greenberg et al. [99] use the spray equation to derive a set of section equations that are consistent with Tambour [26], and again, special attention is made to the form of the vaporization and coalescence source terms. It should be emphasized that no averaging procedure is applied to the equations derived by Greenberg et al. , and as such, the length scales over which the method is valid are questionable. No consideration is given the particle phase dispersion due to the carrier phase turbulence.

Domelevo [101] further develops the approach of Tambour to show that by using the spray equation approach, it is possible to write precise schemes for the evaporation terms. Two significant weaknesses of the method presented by Domelevo [101] are (1) the assumption that there is no relative velocity between the particles and the carrier phase and (2) the method does not account for turbulent particle dispersion. While these assumptions greatly simplify the analysis required, they are also prohibitive if extending the method to more complicated flows of greater general interest.

Laurent and Massot [102] and Laurent et al. [103] again derive a set of sectional transport equations from the "Boltzman" like spray equation and account for Stokes

---

[3] A derivation of Eq. 3.44 can be found in Subramaniam [100]

drag for the particles. Laurent and coworkers make the valid point that by starting from the spray equation allows the inclusion of complex particle phenomena at the "kinetic" level, which is a more natural level to introduce the source terms than the Eulerian field description. It is also acknowledged in this work [102, 103] that no account is made for turbulence in the model, and for this reason, it is only valid for laminar flows or in the spirit of a two-fluid direct numerical simulation.

An interesting sectional method is that of Archambault [104] and Archambault and Edwards [105–107]. Again, the derivation starts from the spray equation of Williams [97] and a set of transport equations are derived for the first three moments of the particle diameter and the particle velocity. It is observed by Archambault that the equations obtained differ from the conventional form in terms of the convection term, which involves a mixed particle velocity moment (e.g. $\langle \phi_p u_{p,i} \rangle$). It is unclear as to why Archambault does not expand this term in the conventional manner

$$\langle \phi_p u_{p,i} \rangle = \left\langle \left( \overline{\phi}_p + \phi'_p \right) \left( \overline{u}_{p,i} + u'_{p,i} \right) \right\rangle = \langle \phi_p \rangle \langle u_{p,i} \rangle + \left\langle \phi'_p u'_{p,i} \right\rangle, \qquad (3.45)$$

and no account is made for the turbulent dispersion of the particulate phase. Archambault decides to abandon the moment equations for the particle diameter due to difficulties encountered with the particle drag terms. These difficulties can be highlighted if the drag term $\langle A_{p,i} \rangle$ for a poly-dispersed spray is considered under Stokes drag

$$\langle A_{p,i} \rangle = -18 \frac{\rho_f \nu_f}{\rho_p} \left[ \left\langle \frac{u_{p,i}}{\phi_p^2} \right\rangle - \left\langle \frac{u_{f,i}}{\phi_p^2} \right\rangle \right], \qquad (3.46)$$

and additionally

$$\langle A_{p,i} u_{p,j} \rangle = -18 \frac{\rho_f \nu_f}{\rho_p} \left[ \left\langle \frac{u_{p,i} u_{p,j}}{\phi_p^2} \right\rangle - \left\langle \frac{u_{f,i} u_{p,j}}{\phi_p^2} \right\rangle \right], \qquad (3.47)$$

which highlight the need to find closures for the mixed moments such as $\left\langle \frac{u_{p,i} u_{f,j}}{\phi_p^2} \right\rangle$.

Instead Archambault chooses to use the sectional approach with transport equations for the first and second moments of particle velocity in each sectional bin. Once again, the method neglects to account for turbulent dispersion mechanism of the particulate phase, and the carrier phase is modeled using an eddy viscosity assumption which is modified to account for the influence of the droplets on the carrier phase. The interesting feature of this work is the closure method proposed for the mixed moment advection term. Because the transport equation does not follow a standard convection-diffusion form, Archambault proposes use of the *maximum entropy principle* [108] to reconstruct the form of the underlying particle PDF from the available moments, and from this, the mixed moment can be calculated. Essentially, the maximum entropy method assures that the PDF of least bias is generated for a given set of input constraints. Details of this method in addition to other more advanced methods will be discussed in Chap. 5.

### 3.3.2 Non-sectional Models

Non-sectional methods are sparse within the literature. However, two notable contributions are discussed here. Firstly, White and Hounslow [109] develop a method for modeling droplet size distributions in poly-dispersed wet steam flows. A control volume approach is used to derive an integral equation analogous to the spray equation of Williams [97] for the particle population. Only convective transport mechanisms are considered, and it is also assumed that there is no relative velocity (and hence no momentum transfer) between the phases. By appropriate integration of the spray equation, White and Hounslow obtain a set of transport equations for the moments of the droplet size distribution. Particular attention is focussed on obtaining closure for the droplet growth source in terms of the available integer moments. While no turbulent dispersion mechanisms are accounted for, White and Hounslow report good agreement with experimental results, with a significant reduction in computation times relative to the traditional sectional methods.

Beck [110] and Beck and Watkins [111–113] present a rather unique approach to modeling poly-dispersed sprays without resorting to sectional methods. Beck derives a set of equations for various moments of the particle size distribution which are convected by a "moment averaged" velocity. For example, the $\left\langle \phi_p^3 \right\rangle$ moment (related to the droplet mass) is convected by a mass averaged velocity. In this way, each size moment is convected with a different velocity field. Upon closer examination of Beck's derivation, it becomes clear that the moment averaged velocity is equivalent to the mixed moment $\dfrac{\left\langle \phi_p^n u_{p,i} \right\rangle}{\left\langle \phi_p^n \right\rangle}$, and actually, this model holds many similarities with the Archambault model [104]. The carrier phase is modeled using a simple eddy viscosity method of Melville and Bray [114], and the same model is employed to model the droplet phase "Moment averaged velocity" and claim good results for a number of test cases [105, 107].

The extensive paper of Minier and Peirano [92] is concerned primarily with stochastic Lagrangian PDF approaches to modeling dispersed particulate flows. However, Minier and Peirano do extend the Lagrangian formalism to demonstrate an equivalence with a set Eulerian field equations. In this extension, a set of transport equations for moments of the particle diameter distribution are derived up to $\left\langle \phi_p'^2 \right\rangle$, but no attempt is made to obtain closure in this work. This formalism provides a foundation for deriving a non-sectional poly-dispersed model of a similar type to the Simonin model [88, 89], and this is explored further in the following chapters.

## 3.4 Summary

The purpose of this chapter was to provide a review of existing Eulerian methods for modeling mono and polydispersed particulate flows. The general conclusions resulting from this review can be summarized as follows:

- The family of RANS methods for particulate flows requires two averaging procedures—firstly to obtain an instantaneous Eulerian description, then a second average to derive a set of macroscopic transport equations for the particulate phase.
- The macroscopic transport equations require an additional model to account for the particle Reynolds stresses. Two methods have been employed—turbulent viscosity methods and second moment closures. The validity of the turbulent viscosity method is questionable given the anisotropic nature of the particulate phase for inertial particles. Conversely, the second moment methods have proved successful in capturing dispersion mechanisms such as inertia and crossing trajectory effects.
- Various complex mathematical techniques have been utilized to derive and close the kinetic-type PDF equation for the particulate phase. The PDF methods require only one ensemble average in contrast to the double averaging for RANS methods.
- The PDF method provides exact analytical solutions for an Eulerian description of the particulate phase including particle history effects; however, these solutions are complex and restrictive in their applicability.
- Inclusion of particle diameter in the PDF phase space density function would allow consideration of polydispersed flows; however, this additional variable would necessitate an additional closure for an additional phase space diffusion term.
- PDF methods based on the Langevin equation have proved successful in capturing the mechanisms of particle dispersion, while also providing a method with the flexibility necessary for viable engineering models. The Langevin approach facilitates particle closure models at the natural Lagrangian level. To date, such models have not been extended to polydispersed flows models without sectioning.
- Several sectional methods have been proposed for modeling particle size distributions. However, these methods suffer from several drawbacks such as the assumption of zero relative velocity between phases and arbitrary or lack of turbulence models.
- A sectional approach based on a second moment model would require particle Reynolds stress transport equations and fluid–particle covariance equations for all size classes. This represents a large increase in the complexity of an already complex system of equations.
- The work toward non-sectional methods has been small, but the potential of these methods has been demonstrated. Nevertheless, these methods have also suffered from similar drawbacks as the sectional methods. To date, a formal basis for these methods has been lacking and if they are to become widespread, it will be necessary to introduce a formal framework.

The remit of this monograph is to provide a systematic method for the development of a non-sectional transport model for polydispersed particulate flows in engineering applications. In order to achieve this goal, the Langevin PDF method would seem to possess the most potential for this purpose given the following reasons:

- The method appears most capable of capturing the anisotropy of the particulate phase and particle dispersion due to inertia and crossing trajectory effects.
- Derivation of the Eulerian transport model from a Lagrangian foundation allows natural addition of new submodels (e.g. evaporation) and guarantees a realizable model at Eulerian level.
- Derivation of a non-sectional model follows naturally from the conservation equation for the PDF in phase space. In addition, the level at which the model is closed can be selected to reduce or increase the complexity of the model.

Having identified the PDF method as the most suitable framework from which to derive the polydispersed model, in the following chapters, a general second moment polydispersed model without size class discretization based on the PDF Langevin method will be derived. The nature and difficulties of the closure issues for the non-integer and third-order moments will then be discussed.

# References

1. Harlow FH, Amsden AA (1975) Numerical calculation of multiphase fluid flow. J Comp Phys 17:19–52
2. Jackson R (1997) Locally averaged equations of motion for a mixture of identical spherical particles and a Newtonian fluid. Chem Eng Sci 52:2457–2469
3. Zhang DZ, Prosperetti A (1997) Momentum and energy equations for disperse two-phase flows and their closure for dilute suspensions. Int J Multiphase Flow 23:425–453
4. Zhang DZ, Prosperetti A (1994) Averaged equations for inviscid disperse two-phase flow. J Fluid Mech 267:185–219
5. Crowe CT, Troutt TR, Chung JN (1996) Numerical models for two-phase turbulent flows. Annu Rev Fluid Mech 28:11–43
6. Crowe CT (1982) Review: numerical models for dilute gas-particle flows. J Fluids Eng 104:297–303
7. Hinze JO (1959) Turbulence. McGraw-Hill, New York
8. Druzhinin OA, Elghobashi S (1997) Direct numerical simulations of bubble-laden turbulent flows using the two-fluid formulation. Phys Fluids 10:685–697
9. Anderson TB, Jackson R (1967) A fluid mechanical description of fluidized beds. Ind Eng Chem Fundam 6:527–539
10. Vernier P, Delhaye JM (1968) General two-phase flow equations applied to the thermodynamics of boiling water nuclear reactors. Energ Primaire 4:5–46
11. Drew DA (1971) Average field equations for two-phase media. Stud Appl Math 50:133–166
12. Whitaker S (1973) The transport equations for multiphase systems. Chem Eng Sci 28:139–147
13. Ishii M (1975) Thermo-fluid dynamic theory of two-phase flow. Eyrolles, Paris
14. Nigmatulin RI (1979) Spatial averaging in the mechanics of heterogeneous and dispersed systems. Int J Multiphase Flow 5:353–385
15. Batchelor GK (1974) Transport properties of two-phase materials with random structure. Ann Rev Fluid Mech 6:227–255
16. Buyevich YA, Shchelchkova IN (1978) Flow in dense suspensions. Prog Aerosp Sci 18: 121–150

17. Drew DA (1983) Mathematical modelling of two-phase flow. Ann Rev Fluid Mech 15: 261–291
18. Shih TH, Lumley JL (1986) Second-order modelling of particle dispersion in a turbulent flow. J Fluid Mech 163:349–363
19. Mashayek F, Pandya RVR (2003) Analytical description of particle/droplet-laden turbulent flows. Prog Energy Comb Sci 29:329–378
20. Favre A (1976) Turbulence en Mécanique des Fluides, CNRS, chap Équations fondamentales des fluids à masse volumique variable en écoulements turbulents, pp 24–78
21. Jones WP, Launder BE (1972) The prediction of laminarization with a two-equation model of turbulence. Int J Heat Mass Trans 15:301–314
22. Pourahmadi F, Humphrey JAC (1983) Modelling solid-fluid turbulent flows with application to prdicting erosive wear. PhysicoChemical Hydrodyn 4:191–219
23. Elghobashi SE, Abou-Arab T (1983) A two-equation turbulence model for two-phase flows. Phys Fluids 26:931–938
24. Elghobashi SE, Abou-Arab T, Rizk M, Mostafa A (1984) Prediction of the particle-laden jet with a two equation turbulence model. Int J Multiph Flow 10:697–710
25. Mostafa AA, Elghobashi SE (1985) A two-equation turbulence model for jet flows laden with vaporizing droplets. Int J Multiph Flow 11:515–533
26. Tambour Y (1980) A sectional model for evaporation and combustion of sprays of liquid fuels. Israel J Tech 18:47–56
27. Rizk MA, Elghobashi SE (1989) Two-equation turbulence model for dispersed dilute confined two-phase flows. Int J Multiph Flow 15:119–133
28. Chen CP, Wood PE (1986) Turbulence closure modelling of the dilute gas-particle axisymmetric jet. AICh E J 32(1):163–166
29. Taulbee DB, Mashayek F, Barre C (1999) Simulation and Reynolds stress modelling of particle-laden turbulent shear flows. Int J Heat Fluid Flow 20:368–373
30. Mashayek F, Taulbee DB (2002) A four equation model for prediction of gas-solid turbulent flows. Num Heat Trans B 41:95–116
31. Issa RI, Oliveira PJ (1993) Engineering turbulence modelling and experiments 2, Elsevier Science Publishers B. V., chap Modelling of turbulent dispersion in two phase flow jets, pp 947–957
32. Picart A, Berlemont A, Gouesbet G (1986) Modelling and predicting turbulence fields and the dispersion of discrete particles transported by turbulent flows. Int J Multiph Flow 12(2):237–261
33. Gouesbet G, Berlemont A, Picart A (1984) Dispersion of discrete particles by continuous turbulent motion. extensive discussion of tchen's theory, using a two—parameter family of lagrangian correlation functions. Phys Fluids 27:827–837
34. Tchen CM (1947) Mean value and correlation problems connected with the motion of small particles suspended in a turbulent fluid. PhD thesis, Delft University, The Hague
35. Abou-Arab TW, Roco MC (1990) Solid phase contribution in the two-phase turbulence kinetic energy equation. J Fluids Eng 112:351–61
36. Lain S, Aliod R (2000) Study of the Eulerian dispersed phase equations in non-uniform turbulent two-phase flows: discussion and comparison with experiments. Int J Heat Fluid Flow 21:374–380
37. Aliod R, Dopazo C (1990) A statistically conditioned averaging formalism for deriving two-phase flow equations. Part Part Syst Charact 7:191–202
38. Reeks MW (1993) On the constitutive relations for dispersed particles in nonuniform flows I: dispersion in a simple shear flow. Phys Fluids A 5:750–761
39. Zhou LX, Gu HX (2003) A nonlinear $k - \varepsilon - k_p$ two-phase turbulence model. J Fluids Eng 125:191–194
40. Zhou LX (2005) Development of multiphase and reacting turbulence models. Num Heat Trans B 47:179–197
41. Barré C, an DB, Taulbee FM (2001) Statistics in particle-laden plane strain turbulence by direct numerical simulation. Int J Multiph Flow 278:347–378

42. Viollet PL, Simonin O, Olive J, Minier JP (1992) Computational methods in applied sciences. Elsevier Science Publishers B, V., chap Modelling turbulent two-phase flows in industrial equipments

43. Mashayek F (1998) Droplet-turbulence interactions in low-Mach-number homogeneous shear two-phase flows. J Fluid Mech 367:163–203

44. Launder BE, Reece GJ, Rodi W (1975) Progress in the development of a reynolds-stress turbulence closure. J Fluid Mech 68:537–566

45. Simonin O, Deutsch E, Bovin M (1995) Turbulent shear flows 9, Springer-Verlag, chap Large eddy simulation and second-moment closure model of particle fluctuating motion in two phase turbulent shear flows, pp 85–115

46. He J, Simonin O (1993) Non-equilibrium prediction of the particle-phase stress tensor in vertical pneumatic conveying. In: Proceedings of 5th international symposium on gas-solid flows, ASME, pp 253–263

47. Grad H (1949) On the kinetic theory of rarefied gases. Commun Pure Appl Math 2:331–407

48. Bertodano MLD, Lee SJ, Lahey RT, Drew DA (1990) The prediction of two-phase flow turbulence and phase distribution phenomena using a reynolds stress model. J Fluids Eng 112:107–113

49. Lain S, Aliod R (2003) Discussion on second-order dispersed phase Eulerian equations applied to turbulent particle-laden jet flows. Chem Eng Sci 58:4527–4535

50. Rodi WA (1976) A new algebraic relation for calculating the Reynolds stresses. ZAMM 56:219–221

51. Pope SB (1975) A more general effective-viscosity hypothesis. J Fluid Mech 72:331–340

52. Taulbee DB (1992) An improved algebraic Reynolds stress model and corresponding nonlinear stress model. Phys Fluids A 4(11):2555–2561

53. Gatski TB, Speziale CG (1993) On explicit algebraic stress models for complex turbulent flows. J Fluid Mech 254:59–78

54. Mashayek F, Taulbee DB (2002) Turbulent gas-solid flows, part II: explicit algebraic models. Numer Heat Transfer Part B 41:31–52

55. Buyevich YA (1971) Statistical hydromechanics of disperse systems. part 1. Physical background and general equations. J Fluid Mech 49:489–507

56. Buyevich YA (1972) Statistical hydromechanics of disperse systems. part 2. Solution of the kinetic equation for suspended particles. J Fluid Mech 52:345–355

57. Buyevich YA (1972) Statistical hydromechanics of disperse systems. part 3. Pseudo-turbulent structure of homogeneous suspensions. J Fluid Mech 56:313–336

58. Reeks MR (1980) Eulerian direct interaction applied to the statistical motion of particles in a turbulent fluid. J Fluid Mech 97:569–590

59. Reeks MW (1991) On a kinetic equation for the transport of particles in turbulent flows. Phys Fluids A 3:446–456

60. Reeks MW (1992) On the continuum equations for dispersed particles in nonuniform flows. Phys Fluids A 4:1290–1303

61. Gambosi TI (1994) Gaskinetic theory. Cambridge University Press, Cambridge

62. Subramaniam S (2000) Statistical representation of a spray as a point process. Phys Fluids 12:2413–2431

63. Hyland KE, McKee S, Reeks MW (1999) Derivations of a PDF kinetic equation for the transport of particles in turbulent flows. J Phys A: Math Gen 32:6169–6190

64. Gardiner CW (1985) Handbook of stochastic methods for physics, chemistry and the natural sciences. Springer, Berlin (Springer series in synergetics)

65. Kraichnan RH (1958) The structure of isotropic turbulence at very high Reynolds numbers. J Fluid Mech 5:497–543

66. Kraichnan RH (1965) Lagrangian-history closure approximation for turbulence. Phys Fluids 8:575–598

67. Hyland KE, McKee S, Reeks MW (1999) Exact analytical solutions to turbulent particle flows. Phys Fluids 11:1249–1303

68. Derevich IV (2000) Statistical modelling of mass transfer in turbulent two-phase dispersed flows—1 model development. Int J Heat Mass Trans 43:3709–3723
69. Derevich IV (2000) Statistical modelling of mass transfer in turbulent two-phase dispersed flows—2 calculation results. Int J Heat Mass Trans 43:3725–3734
70. Zaichik LI (1999) A statistical model of particle transport and heat transfer in turbulent shear flows. Phys Fluids 11:1521–1534
71. Pandya RVR, Mashayek F (2001) Probability density function modelling of evaporating droplets dispersed in isotropic turbulence. AIAA J 39:1909–1915
72. Porta A, Voth G, Crawford M, Alexander J, Bodenschatz E (2001) Fluid particle accelerations in fully developed turbulence. Nature 409:1017–1019
73. Zaichik LI, Simonin O, Alipchenkov VM (2003) Two statistical models for predicting collision rates of inertial particles in homogenous isotropic turbulence. Phys Fluids 15:2995–3005
74. Zaichik LI, Alipchenkov VM (2001) A statistical model for transport and deposition of high-inertia colliding particles in turbulent flow. Int J Heat Fluid Flow 22:365–371
75. Zaichik LI, Pershukov VA, Kozelev MV, Vinberg AA (1997) Modelling of dynamics, heat transfer, and combustion in two-phase turbulent flows: 1. Isothermal flows. Eperimental Therm Fluid Sci 15:291–310
76. Zaichik LI, Pershukov VA, Kozelev MV, Vinberg AA (1997) Modelling of dynamics, heat transfer, and combustion in two-phase turbulent flows: 2. Flows with heat transfer and combustion. Experimental Therm Fluid Sci 15:311–322
77. Zaichik LI, Alipchenkov VM (2005) Statistical models for predicting particle dispersion and preferential concentration in turbulent flows. Int J Heat Fluid Flow 26:416–430
78. Kampen NGV (1974) A cumulant expansion for stochastic linear differential equations. I. Physica 74:215–238
79. Kampen NGV (1974) A cumulant expansion for stochastic linear differential equations. II. Physica 74:239–247
80. Pozorski J, Minier JP (2001) Probability density function modelling of dispersed two-phase turbulent flows. Phys Rev E 59:855–863
81. Pandya RVR, Mashayek F (2003) Non-isothermal dispersed phase of particles in turbulent flow. J Fluid Mech 475:205–245
82. Simonin O, Deutsch E, Bovin M (1993) Large eddy simulation and second-moment closure model of particle fluctuating motion in two-phase turbulent shear flows. In: Turbulent shear flows 9
83. Simonin O, Deutsch E, Minier JP (1993) Eulerian prediction of the fluid/particle correlated motion in turbulent two-phase flows. App Sci Res 51:275–283
84. Simonin O (1991) Second-moment prediction of dispersed phase turbulence in particle-laden flows. In: Eighth symposium on turbulent shear flows
85. Csanady GT (1963) Turbulent diffusion of heavy particles in the atmosphere. J Atmos Scie 20:201–208
86. Wells MR, Stock DE (1983) The effects of crossing trajectories on the dispersion of particles in a turbulent flow. J Fluid Mech 136:31–62
87. Haworth DC, Pope SB (1986) A generalized Langevin model for turbulent flows. Phys Fluids 29:387–405
88. Simonin O (2000) Statistical and continuum modelling of turbulent reactive particulate flows. part 1: theoretical derivation of dispersed phase Eulerian modelling from probability density function kinetic equation. In: Lecure series, Von-Karman Institute for Fluid Dynamics
89. Simonin O (2000) Statistical and continuum modelling of turbulent reactive particulate flows. part 2: application of a two-phase second-moment transport model for prediction of turbulent gas-particle flows. In: Lecure series, Von-Karman Institute for Fluid Dynamics
90. Monin AS, Yaglom AM (1971) Statistical fluid mechanics. MIT Press, Cambridge
91. Singh K, Squires KD, Simonin O (2004) Evaluation using an LES database of constitutive relations for fluid-particle velocity correlations in fully developed gas-particle channel flow. In: International conference on multiphase flow

92. Minier JP, Peirano E (2001) The pdf approach to turbulent polydispersed two-phase ows. Phys Rep 352:1–214

93. Peirano E, Minier JP (2002), Probabilistic formalism and heirarchy of models for polydispersed turbulent two-phase flows. Phys Rev E 65:463, 011–4630, 118

94. Elperin T, Kleeorin N, Rogachevshii I (1998) Anomalous scalings for fluctuations of inertial particles concentration and large-scale dynamics. Phys Rev E 58(3):3113–3124

95. Elperin T, Kleeorin N, Rogachevshii I (2000) Passive scalar transport in a random flow with finite renewal time: mean field equations. Phys Rev E 61(3):2617–2625

96. Elperin T, Kleeorin N (1996) Turbulent thermal diffusion of small inertial particles. Phys Rev Let 76(2):224–227

97. Williams FA (1958) Spray combustion and atomization. Phys Fluids 1:541–545

98. Williams FA (1985) Combustion theory: the theory of chemically reacting flow systems. Benjamin Cummings, Menlo Park

99. Greenberg JB, Silverman I, Tambour Y (1992) On the origins of spray sectional conservation equations. Combust Flame 93:90–96

100. Subramaniam S (2001) Statistical modelling of sprays using the droplet distribution function. Phys Fluids 13:624–642

101. Domelevo K (2001) The kinetic sectional approach for noncolliding evaporating sprays. Atomization Sprays 11:291–303

102. Laurent F, Massot M (2001) Multi-fluid modelling of laminar polydispersed spray flames: origin, assumptions and comparrison of sectional and sampling methods. Combust Theory Model 5:537–572

103. Laurent F, Massot M, Villedieu P (2004) Eulerian multi-fluid modelling for the numerical simulation of coalescence in polydisperse dense liquid sprays. J Comp Phys 194:505–543

104. Archambault MR (1999) A maximum entropy moment closure approach to modelling the evolution of spray flows. PhD thesis, Stanford

105. Archambault MR, Edwards CF (2000) Computation of spray dynamics by direct solution of moment transport equations—inclusion of nonlinear momentum exchange. In: Eighth international conference on liquid atomization and spray systems

106. Archambault MR, Edwards CF, McCormack RW (2003) Computation of spray dynamics by moment transport equations I: theory and development. Atomization Sprays 13:63–87

107. Archambault MR, Edwards CF, McCormack RW (2003) Computation of spray dynamics by moment transport equations II: application to quasi-one dimensional spray. Atomization Sprays 13:89–115

108. Jaynes ET (1957) Information theory and statistical mechanics. Phys Rev 106:620–630

109. White AJ, Hounslow MJ (2000) Modelling droplet size distributions in polydispersed wet-steam flows. Int J Heat Mass Trans 43:1873–1884

110. Beck JC (2000) Computational modelling of polydisperse sprays without segregation into droplet size classes. PhD thesis, UMIST

111. Beck JC, Watkins AP (1999) Spray modelling using the moments of the droplet size distribution. In: ILASS-Europe

112. Beck JC, Watkins AP (2003) The droplet number moments approach to spray modelling: the development of heat and mass transfer sub-models. Int J Heat Fluid Flow 24:242–259

113. Beck J, Watkins AP (2003) Simulation of water and other non-fuel sprays using a new spray model. Atomization Sprays 13:1–26

114. Melville WK, Bray KNC (1979) A model of the two-phase turbulent jet. Int J Heat Mass Tran 22:647–656

# Chapter 4
# A Poly-dispersed EE Model

In this chapter, first a brief introduction to kinetic theory is provided to demonstrate the terminology which the authors believe helps in grasping the ideas and then, the basic definitions for the fluid–particle system are provided. In this chapter, the stochastic framework is used to develop a consistent second-moment model for poly-sized particulate flows, in keeping with the ideas presented in Minier and Peirano [1]. The start point for the model derivation is the specification of the SDEs that describe the joint evolution of the particle and fluid element pair. In the most general case, these equations must account for interphase mechanisms such as mass and momentum transfer. These terms can be explicitly accounted for at the SDE mesoscopic level, which is a clear motivation for using this Lagrangian description in favour of the more traditional Eulerian method. In addition to this advantage, the stochastic Lagrangian approach guarantees a realizable model (providing real and bounded SDE coefficients) at the macroscopic Eulerian level [2]. In the following section, a joint particle–fluid SDE description is proposed and justified in terms of several key physical processes such as crossing trajectory effects [3, 4] and modification of fluid turbulence due to the presence of particles [5]. Following this, the full second-moment model is derived in general terms starting from the SDE mesoscopic description. Finally, closure issues are highlighted and discussed which provides motivation for the following chapters.

## 4.1 Single-Particle Motion

### 4.1.1 Molecular Particle Motion

The trajectory of a single particle surrounded by other smaller particles was first studied rigorously by Einstein. The motion of such a particle consists of a large number of instantaneous changes in velocity and if we assume that the frequency of molecular collisions is very high and the motion of the particle in different time intervals is independent as long as the intervals are not chosen too small, then such particle undergoes a stochastic Wiener process, which is also known as Brownian motion. Properties of the Wiener process and Brownian motion were discussed in detail in Sect. 2.4.

### 4.1.2 Macroscopic Particle Motion

Historically, the equation of motion of a single macroscopic particle was studied by Stokes and he derived a well-known formula for drag coefficient, based on this study [6]. To take into account forces other than the drag, a good starting point would be the Basset–Boussinesq–Oseen (BBO) equation [7–9]. This equation later revised by Maxey and Riley [10] who added to it an extra term to account for non-uniformity of velocity field and the additional Basset history force resulting in [8]:

$$m_p \frac{\mathrm{d}U_{p,i}}{\mathrm{d}t} = F_{p,i} + F_{\mathrm{vis},i} + F_{\mathrm{D},i} + F_{\mathrm{AM},i} + F_{\mathrm{H},i} + F_{\mathrm{G},i}, \qquad (4.1)$$

where $F_{p,i}$, $F_{\mathrm{vis},i}$, $F_{\mathrm{D},i}$, $F_{\mathrm{AM},i}$, $F_{L,i}$, $F_{\mathrm{H},i}$ and $F_{\mathrm{G},i}$ are pressure, viscous, drag, added mass, Basset history, and gravity force. Viscous force, which can be ignored relative to the pressure force [8], can be written as $V_p \frac{\partial \tau_{ij}}{\partial x_j}$, $V_p$ being the particle volume and $\tau_{ij}$ the fluid shear stress tensor. Pressure, viscous and gravity forces can also be interpreted as buoyancy and fluid acceleration ($F_{B,i}$, $F_{A,i}$) by recalling the Navier–Stokes equations. In gas–particle flow, where the ratio of material densities, i.e $\rho_f / \rho_d$ where $\rho_f$ and $\rho_d$ are fluid and particle densities, respectively, is usually in order of $10^{-3}$ [9]. When a particle accelerates in a fluid, the surrounding fluid also accelerates correspondingly at the expense of the work done by the body resulting in a force, $F_{\mathrm{AM},i}$, which should be added to equation of motion of the particle. Basset history force is due to the delay in formation of the boundary layer around the particle as the velocity changes with time [8, 9]. Both Basset history force and the added mass force can be neglected again if the density ratios $\rho_f / \rho_d < 10^{-3}$ and also if the frequency of oscillation in the stream is not too high, i.e. $Re_\omega < 36$, where $Re_\omega$ is the Reynolds number formed by the frequency of oscillations ($\omega$) in the flow and the particle diameter ($D_p$) [11–13]. These equations are the basis of the EL and point-particle DNS simulations, c.f. Sect. 1.2.

Another important force acting on the particle not included in the original Maxey and Riley equation [10] is the Saffman lift force, and the full form was originally derived by Auton [14, 15]. Particles rising in a non-uniform flow field experience a lift force due to the vorticity or the shear in the flow field. Auton [14, 15] showed that this lift force depends on the vector product of the slip velocity and the curl of the fluid velocity, resulting in a force that is perpendicular to both the direction of the slip velocity and the direction of the fluid vorticity field. There are other forces that can be added to this equation depending on the state of the flow. Coulomb forces due to electrostatic effects [16] and Magnus force [17] due to particle rotation are some examples of these other forces.

From the above discussion, it is evident that under assumption of small density ratio and low frequency oscillating flow and in absence of other effects, the equation can be significantly simplified to:

$$m_p \frac{dU_{p,i}}{dt} = F_{D,i},$$ (4.2)

with drag force given by, see for example [9, 18]:

$$F_{D,i} = -3\pi D_p \mu_f \left( U_{p,i} - U_{f,i}|_{X_{p,i}} - \frac{D_p^2}{24} \frac{\partial^2 U_{f,i}}{\partial x_j \partial x_j}|_{X_{p,i}} \right)$$ (4.3)

where $D_p, \mu_f, U_{p,i}$ and $U_{f,i}$ are particle diameter, fluid viscosity, particle velocity and fluid velocity and a subscript $X_{p,i}$ means the property evaluated at particle position.

As discussed in Sect. 1.2, fully resolved direct numerical simulation of problems of practical interest is not feasible and a point-particle approach with suitable models for drag and lift coefficients is an appropriate starting point for a practical model. Regardless of the number of forces retained in the equation when using a point-particle approach, the forces acting on the particle are calculated based on the assumption of an undisturbed ambient flow that would exist in absence of the particle phase, see for example [19–21] and the references therein. This assumption is only valid if the particle diameter is much smaller than Kolmogorov length scale, $D_p \ll \eta$; however, if for example a LES approach is used for modelling turbulence, this restriction is relaxed to diameters smaller than the filtered scales, i.e $D_p \ll \xi$, [22–24]. Additional complications arise when trying to feed back the forces to the fluid phase, i.e two-way coupling. Theoretically, accounting for two-way coupling is straightforward as this force is the negative of the force acting on the particle, however, severe restrictions and complications arise due to the original assumption of the undisturbed fluid field [19].

## 4.2 Introduction to PDF Method Through Kinetic Theory

The PDF approach discussed in the following sections is similar to kinetic theory of gases (KTG) [25, 26] and a general discussion of the kinetic theory, the Boltzmann equation and also its direct application to a dense particle phase resulting in the kinetic theory of granular flows is productive. Despite the similarities, one should bear in mind that gas kinetic theory and the method discussed in this review are only similar in terms of methodology and this section introduces the methodology using the kinetic theory in a more physically sensible way. In statistical mechanics, it is believed that fluid properties can be understood from detailed knowledge of the state of all atoms or molecules that constitute the fluid. These states develop based on classical laws of mechanics which for the system at hand can be written as [27]:

$$\frac{\mathrm{d}X_{k,i}}{\mathrm{d}t} = U_{k,i}$$

$$m_k \frac{\mathrm{d}U_{k,i}}{\mathrm{d}t} = f_{k,i} + \sum_{k'=1}^{N} f_{k'k,i} \tag{4.4}$$

The initial conditions for Eq. (4.4) are:

$$X_{k,i}(t = 0) = X_{k_0,i}$$

$$U_{k,i}(t = 0) = U_{k_0,i} \quad k = 1 \ldots N, \ i = 1 \ldots 3 \tag{4.5}$$

In Eq. (4.4), $X_{k,i}$ is the coordinate of the '$k$th' molecule in $i$ direction and $U$ is its velocity. $f_{k,i}$ is the sum of all forces acting on the molecule other than collisions, and $f_{kk',i}$ is the collision force between each $k$ and $k'$ pair. Also, note that in Eq. (4.4), we are considering all the molecules and interactions, so neither $f_{k,i}$ nor $f_{kk',i}$ are stochastic. There are several issues associated with this approach. Firstly, the number of ODEs to be solved is $6N$ and for a gas at standard conditions, $N$ is approximately $10^{20}$ and obviously solving this number of equations is an impossible task. Secondly, to measure the initial position and velocity of a huge number of particles, i.e. molecules or atoms, at $t = 0$ to provide the initial conditions. Selection of a valid averaging which should be performed in a volume is small compared to the large scales of the macroscopic quantities, such as pressure, density or temperature, yet large enough to contain enough number of molecules to get reliable statistics of the macroscopic quantities is another issue. In this limit, fluctuations are inevitable and constitutive relations are needed [27, 28].

A macroscopic approach, which is considering the phase as a continuum and deriving the evolution equation of the fluid properties under the assumption of continuity, also becomes invalid as the molecular distances become comparable to that of the large scales of the flow; therefore, the continuity assumption becomes invalid. The Boltzmann approach overcomes these issues by defining a $6N$-dimensional space where coordinates are $3N$ components of $N$ position vectors and $3N$ components of $N$ velocity vectors. In this space, the state of the system, if known, is presented by a point with $6N$ coordinate values. Using this phase space, we define the distribution $\mathscr{P}(\mathbf{x}_1, \ldots, \mathbf{x}_N; \mathbf{u}_1, \ldots, \mathbf{u}_N; t)$ such that the quantity $\mathscr{P}\mathrm{d}\Gamma$, where $\mathrm{d}\Gamma$ is the infinitesimal element of phase space spanned by the coordinates and velocity of all particles defined by $\prod_{k=1}^{N} \mathrm{d}\mathbf{u}_k \mathrm{d}\mathbf{x}_k$, is the probability that at time $t$ particles with coordinate $(\mathbf{x}_1, \ldots, \mathbf{x}_N, \mathbf{u}_1, \ldots, \mathbf{u}_N)$ can be found in the volume defined by $\mathrm{d}\Gamma$. Obviously, this distribution is far from practical to work with, so restrict ourselves to the case of a single-particle distribution $\mathscr{P}(\mathbf{x}; \mathbf{u}; t)$, such that the total number of molecules can be defined by:

$$N = \int \mathscr{P}(\mathbf{x}, \mathbf{u}, t)\mathrm{d}\mathbf{x}\mathrm{d}\mathbf{u} \tag{4.6}$$

Cercignani [28, 29] derived a rigorous formula for the evolution of the whole distribution and showed that, for example the PDF evolution of a single particle depends on the two-particle distribution functions and so on. This creates a hierarchy of integrals which are known as BBGKY hierarchy and can also be interpreted as a consequence of reducing an n-point PDF to a single-point PDF (contracting a PDF). However, a simple intuitive derivation of the single-particle evolution equation can be performed by considering a gas subject to an external force $m A_i$ which can depend on position $X_{p,i}$ but not on velocity $U_{p,i}$ [25]. As stated before, the number of molecules in time $t$ at position $X_{p,i}$ with velocity $U_{p,i}$ is

$$\mathscr{P}(\mathbf{x}, \mathbf{u}, t) d\mathbf{x} d\mathbf{u} \tag{4.7}$$

And at $t + dt$, the number of molecules with velocity $u_{p,i} + A_i dt$ at position $x_{p,i} + u_{p,i} dt$ neglecting collisions would be:

$$\mathscr{P}(\mathbf{x} + \mathbf{u}dt, \mathbf{u} + \mathbf{A}dt, t + dt) d\mathbf{x} d\mathbf{u} \tag{4.8}$$

Then, denoting the net rate of change of molecules by encounters as $\frac{\partial_e \mathscr{P}}{\partial t}$, subtracting Eq. (4.8) from Eq. (4.7) and dividing by $d\mathbf{u}d\mathbf{x}dt$, we get [25]:

$$\frac{\partial \mathscr{P}}{\partial t} + \dot{z}_{p,i} \frac{\partial \mathscr{P}}{\partial z_{p,i}} = \frac{\partial_e \mathscr{P}}{\partial t} \tag{4.9}$$

where $z$ is the state vector containing both velocities and positions, i.e. $i = 1 \ldots 6$ for a single particle, and $\dot{z}_{p,i}$ is the time derivative of the state vector corresponding to the velocities and external forces acting on the particle. Equation (4.9) without the encounter term is the Liouville equation which simply states that $\mathscr{P}$ takes at time $t$ and at point $z$ the value it took at time $t_0$ and point $z_0$ carried to its current state by the motion described by Eq. (4.9).

Having the probability density and its evolution in time, we can find the transport equation of the macroscopic measurable quantities of interest as averages. This can be done by starting from the Boltzmann equation, Eq. (4.9), multiplying both sides by an arbitrary quantity $\Psi$ and integrating over all velocities using general divergence theorem transformations which results in Maxwell's equation [25, 30], see [31–33] for a derivation of Navier–Stokes equations from the Boltzmann equation.

To use the same approach for the particulate flows, see [34–36], the integral of form $\int \frac{\partial_e \mathscr{P}}{\partial t} du_p$ appearing on the RHS of the Boltzmann equation, Eq. (4.9), should be approximated for different type of particulate flows. Gidaspow and Neri [30, 37] approximate this term for dilute and dense particulate flows resulting in kinetic theory of granular flow (KTGF). Although Gidaspow and Neri obtain reasonable results, the applicability of the method specially to dilute systems is questionable [38], also inclusion of turbulence effects which is far more important in dilute systems than the particle–particle collisions is not straightforward.

## 4.3 PDF Evolution Equation

We have demonstrated the weak equivalence, meaning Lagrangian equations contain more information, between Lagrangian stochastic description and evolution of underlying PDF [39]. The starting point for the model derivation is the specification of the SDEs that describe the joint evolution of the particle and fluid element pair, in which at the most general case should contain the interphase mechanisms such as mass and momentum transfer. One advantage of the current approach is the explicit addition of these terms to the equations and the other is that a realizable model is guaranteed at the macroscopic Eulerian level [40]. In this section, we write the Lagrangian equations and derive the moment equation of a general property by integrating the PDF evolution equation. We also discuss inclusion of key physical processes such as crossing trajectory effect [3, 41] and turbulent field modification due to the presence of the particles [5].

### *4.3.1 Deterministic One-Point, Two-Particle Description*

A deterministic fluid–particle system can be defined by state vector

$$\mathbf{Z}_{\text{fp}}^{+} = \left( X_{f,i}^{+},\, U_{f,i}^{+},\, \Psi_{f,m}^{+},\, X_{\text{p},i}^{+},\, U_{\text{p},i}^{+},\, \Psi_{\text{p},n}^{+} \right), \tag{4.10}$$

or alternatively by its corresponding phase space vector

$$\mathbf{z}_{\text{fp}}^{+} = \left( y_{f,i}^{+},\, u_{f,i}^{+},\, \psi_{f,m}^{+},\, y_{\text{p},i}^{+},\, u_{\text{p},i}^{+},\, \psi_{\text{p},n}^{+} \right). \tag{4.11}$$

Distinction is made between fluid/particle position in a Lagrangian reference frame $(y_{k,i})$ and an Eulerian frame $(x_{k,i})$ to emphasize the transition from the Lagrangian mesoscopic description to the Eulerian macroscopic description. One should bear in mind, however, that this is just a reduced description of the real system which contains many particles and many degrees of freedom in the fluid phase but no additional notation is used to indicate this 'reduced' state providing a simplified notation. The consequences of this reduction will become clear when trying to derive the PDF evolution equation from the trajectories of the fluid and solid particles. The fluid trajectories can be defined as:

$$dX_{f,i}^{+} = U_{f,i}^{+} dt \tag{4.12}$$

$$dU_{f,i}^{+} = A_{f,i}^{+} dt + A_{\text{p} \to f,i}^{+} dt, \tag{4.13}$$

where the '+' is used to indicate the deterministic nature of this formulation. $A_{\text{p} \to f,i}^{+}$ is the influence of the particles on the fluid. If the fluid phase is assumed to be isothermal, non-reactive and incompressible, there is no need to write an ODE for $d\Psi_{f,m}$. The fluid acceleration term is given by

$$A_{f,i}^+ = -\rho_f^{-1}\partial P^+/\partial x_i + v_f \partial^2 U_{f,i}^+/(\partial x_j \partial x_j), \qquad (4.14)$$

and inter-phase momentum exchange by

$$A_{p\to f,i}^+ = A_{p\to f,i}^+(\mathbf{Z}_{fp}^+, \langle \mathbf{Z}_{fp}^+ \rangle), \qquad (4.15)$$

which is undefined at this stage. The particle state vector is defined in a similar manner by $\mathbf{Z}_p^+ = \left(X_{p,i}^+, U_{p,i}^+, \Psi_{p,n}^+\right)$ and deterministic trajectories are:

$$dX_{p,i}^+ = U_{p,i}^+ dt \qquad (4.16)$$

$$dU_{p,i}^+ = A_{p,i}^+ dt, \qquad (4.17)$$

where particle acceleration $A_{p,i}^+ = \frac{1}{\tau_p}(U_{s,i}^+ - U_{p,i}^+) + g_i$. $U_{s,i}$ is the fluid velocity at the particle location or 'seen' by the particle. If we further assume at this stage that particles can evaporate isothermally but agglomeration, coalescence and collision effects are negligible, then equation for particle diameter can simply be written by $d\Phi_p^+ = \Theta_p^+ dt$, $\Phi_p^+$ being particle diameter and $\Theta_p^+$ its the evaporation rate. Now using the Liouville equation, Eq. (2.35), the Lagrangian PDF evolution equation for joint fluid–particle system can easily be written which, using Eq. (2.52) and integrating the result using Eq. (2.53), can be expressed in terms of Eulerian MDF as:

$$\frac{\partial \mathscr{F}_{fp}^{E+}}{\partial t} + \frac{\partial}{\partial y_{f,i}^+}\left[v_{f,i}^+ \mathscr{F}_{fp}^{E+}\right] + \frac{\partial}{\partial y_{p,i}^+}\left[v_{p,i}^+ \mathscr{F}_{fp}^{E+}\right]$$

$$= -\frac{\partial}{\partial v_{f,i}^+}\left[\left\langle A_{f,i}^+|\mathbf{Z}_{fp}^+ = \mathbf{z}_{fp}^+\right\rangle \mathscr{F}_{fp}^{E+}\right] - \frac{\partial}{\partial v_{p,i}^+}\left[\left\langle A_{p,i}^+|\mathbf{Z}_{fp}^+ = \mathbf{z}_{fp}^+\right\rangle \mathscr{F}_{fp}^{E+}\right]$$

$$- \frac{\partial}{\partial v_{f,i}^+}\left[\left\langle A_{p\to f,i}^+|\mathbf{Z}_{fp}^+ = \mathbf{z}_{fp}^+\right\rangle \mathscr{F}_{fp}^{E+}\right] - \frac{\partial}{\partial \delta_p^+}\left[\left\langle \Theta_p^+|\mathbf{Z}_{fp}^+ = \mathbf{z}_{fp}^+\right\rangle \mathscr{F}_{fp}^{E+}\right],$$

$$(4.18)$$

which is a two-point, two-particle description of the system. The terms of form $\langle A^+|\mathbf{Z}_{fp}^+ = \mathbf{z}_{fp}^+\rangle$ appear here, as a result of choosing reduced state vector which means to get to this equation another implicit integration of form

$$\int \frac{\partial \mathscr{P}(\mathbf{z};t)}{\partial t}d\mathbf{z}^{n-r} + \int \frac{\partial}{\partial \mathbf{z}}[A(\mathbf{z},t)\mathscr{P}(\mathbf{z}^{n-r}|\mathbf{z}^r;t)\mathscr{P}(\mathbf{z}^r;t)]d\mathbf{z}^{n-r} = 0, \quad (4.19)$$

is needed which after defining

$$\langle A|\mathbf{Z}^r = \mathbf{z}^r \rangle = \int A(\mathbf{z}, t) \mathscr{P}(\mathbf{z}^{n-r}|\mathbf{Z}^r = \mathbf{z}^r; t) \mathrm{d}\mathbf{z}^{n-r}, \qquad (4.20)$$

gives the Liouville equation of the reduced quantity state vector as

$$\frac{\partial \mathscr{P}(\mathbf{z}^r; t)}{\partial t} + \frac{\partial}{\partial \mathbf{z}} [\langle A(\mathbf{z}, t)|\mathbf{Z}^r = \mathbf{z}^r \rangle \, \mathscr{P}(\mathbf{z}^r; t)] = 0 \qquad (4.21)$$

It is also worth mentioning the similarities between Eq. (4.18) and spray equation of Williams [42] used by Archambault [43–45], Demelevo [46] and Tambour [47]. The spray equation without any source terms can be expressed as:

$$\frac{\partial F_{\mathrm{p}}^{E+}}{\partial t} + \frac{\partial}{\partial y_{\mathrm{p},i}^+} \left[ v_{\mathrm{p},i}^+ F_{\mathrm{p}}^{E+} \right] + \frac{\partial}{\partial v_{\mathrm{p},i}^+} \left[ \left\langle A_{\mathrm{p},i}^+ | \mathbf{Z}_{\mathrm{p}}^+ = \mathbf{z}_{\mathrm{p}}^+ \right\rangle F_{\mathrm{p}}^{E+} \right]$$

$$+ \frac{\partial}{\partial \delta_{\mathrm{p}}^+} \left[ \left\langle \Theta_{\mathrm{p}}^+ | \mathbf{Z}_{\mathrm{p}}^+ = \mathbf{z}_{\mathrm{p}}^+ \right\rangle F_{\mathrm{p}}^{E+} \right] = 0. \qquad (4.22)$$

An identical equation can be obtained from Eq. (4.18) by integrating over all fluid variables leading to the spray equation which is a one-point, one-particle description for a two-phase flow meaning that we are only writing the equation for the particle phase and fluid phase PDF equations are ignored at this stage. This comparison highlights the advantages of the statistical approach adopted here and also demonstrates some of the weaknesses of previous models. For example, Eq. (4.18) preserves the coupling between the fluid and particulate phases, whereas in the spray equation method, the fluid description is external throughout. In addition, Eqs. (4.18) and (4.22) pertain to a deterministic systems; in other words, no time averaging has been applied at this stage and addition of turbulent effects is not trivial. However, moving from this deterministic description to a stochastic description which takes into account turbulent effects and eliminates the need for integration of form (4.19) for fluid and particle acceleration terms is straightforward although modelling is still required for fluid–particle interactions. This is done by adding pure Wiener processes to the trajectory equations and will be discussed in detail in the next section.

### 4.3.2  Choice of the Variables in the State Vector

Before discussing the stochastic description of the fluid–particle system, it is necessary to discuss how the state vector variables should be chosen. The choice of variable retained in the state vector is usually done through fast variable elimination technique [39, 48]. In this technique, variables with timescales much smaller than the timescale of the observed phenomena are called fast variables. It is then assumed that these variables reach their stationary states very rapidly and consequently can be expressed as functions of slower modes and a Wiener process to account for their

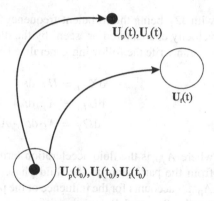

**Fig. 4.1** The particle crossing trajectory effect due to mean velocity between discrete particle and fluid element

$U_p(t), U_s(t)$

$U_f(t)$

$U_p(t_0), U_s(t_0), U_f(t_0)$

stochastic nature [1, 39]. It can be shown using Kolmogorov theory that velocity is a slow variable, while acceleration is a fast variable for a turbulent fluid flow; consequently, position and velocity should be retained in the state vector [40, 49]. However, we also add turbulent frequency to the fluid state vector which will be discussed in Sect. 4.4.1.3.

The extension of the stochastic description to the particle phase is more difficult due to the fact that the fluid velocity measured at the particle position and the particle velocity are random but correlated variables [50] which are influenced by factors such as particle inertia and the so-called crossing trajectory effect [3, 41]. The crossing trajectory effect is caused by relative motion (mean and turbulent) between the fluid and the particle which leads to a reduction of the particle Lagrangian velocity auto-correlation, see Fig. 4.1, with respect to the fluid element [3, 41]. There are several methods to include this effect into the formulation [51–53]; however, these methods result in inconsistencies in the limiting case of very small particle relaxation times specially in non-homogeneous flows [54]. Thus, a better proposal is to include the seen velocity directly into the state vector [54–57] resulting in a state vector which consists of particle position, particle velocity and fluid velocity seen by the particle, or more correctly fluid velocity sampled along the trajectory of the particle.

### 4.3.3 Stochastic One-Point, Two-Particle Description

We first define the fluid–particle state vector

$$\mathbf{Z}_{\mathrm{fp}} = \left( X_{f,i}, U_{f,i}, \Omega_f, X_{\mathrm{p},i}, U_{\mathrm{p},i}, U_{\mathrm{s},i}, \Phi_{\mathrm{p}} \right),  \quad (4.23)$$

with the corresponding phase space vector

$$\mathbf{z}_{\mathrm{fp}} = \left( y_{f,i}, u_{f,i}, \omega_f, y_{\mathrm{p},i}, u_{\mathrm{p},i}, u_{\mathrm{s},i}, \phi_{\mathrm{p}} \right),  \quad (4.24)$$

with $\Omega_f$ being the turbulent frequency along the fluid element trajectory and $U_{s,i}$ velocity encountered or 'seen' by the discrete particle. Then, for the fluid element, we can write the following generalized SDE system:

$$dX_{f,i} = U_{f,i}dt \tag{4.25}$$

$$dU_{f,i} = A_{f,i}dt + A_{p\to f,i}dt + B_{f,ij}dW_{f,j} \tag{4.26}$$

$$d\Omega_f = A_\Omega dt + A_{p\to\Omega}dt + B_\Omega dW_\Omega, \tag{4.27}$$

where $A_{f,i}$ is the fluid acceleration term, $A_{p\to f,i}$ accounts for momentum transfer from the particulate to the fluid phase, $A_\Omega$ is the turbulent frequency drift term, $A_{p\to\Omega}$ accounts for the influence of the particles on the fluid turbulent frequency, and finally, $B_{f,ij}$ and $B_\Omega$ are the diffusion terms corresponding to the stochastic Wiener processes $dW_{f,j}$ and $dW_\Omega$, respectively. As discussed in Sect. 4.3.2, traditionally, only velocity and position vectors are kept in the fluid state vector [58] and while this PDF describes the distribution of velocities, its description of the turbulent motion is seriously deficient [59] in that it provides no information on the length or timescale of the turbulent motion. Thus, in the current model, we add the turbulent frequency to the state vector to get a better description of the turbulent flow field. The discrete particle trajectories considering a non-evaporating dispersed phase, $d\Phi_p = 0$, are as follows

$$dX_{p,i} = U_{p,i}dt \tag{4.28}$$

$$dU_{p,i} = A_{p,i}dt \tag{4.29}$$

$$dU_{s,i} = A_{s,i}dt + A_{p\to s,i}dt + B_{s,ij}dW_{s,j}, \tag{4.30}$$

where the particle drift coefficient $A_{p,i}$ is taken as the particle acceleration due to aerodynamic drag, $A_{s,i}$ is the drift coefficient for the fluid element sampled or seen by the particle, and $A_{p\to s,i}$ accounts for the momentum transfer between the particulate and sampled velocity phases. Momentum transfer terms between phases $A_{p\to s,i}$, $A_{p\to f,i}$ and $A_{p\to\Omega}$ are defined by [1]:

$$A_{p\to s,i} = -\frac{\alpha_p\rho_p}{\alpha_f\rho_f}\left(\frac{U_{s,i} - U_{p,i}}{\tau_p}\right), \quad A_{p\to f} = -\frac{\alpha_p\rho_p}{\alpha_f\rho_f}\left(\frac{U_{s,i} - U_{p,i}}{\tau_p}\right),$$

$$A_{p\to\Omega} = -\frac{\rho_p U_{f,i}}{\rho_f}\left(\frac{U_{s,i} - U_{p,i}}{\tau_p}\right)D_\Omega$$

where

$$D_\Omega = E_\Omega \exp\left[\left(\frac{\Omega - \tau_p^{-1}}{\sigma_\Omega}\right)^2\right]. \tag{4.31}$$

In Eq. (4.31), $\alpha_p \rho_p$ is the expected particle mass at position $x_i$, $\alpha_f \rho_f$ is the expected mass of fluid, and $\tau_p$ is the particle timescale. The terms $E_\Omega$ and $\sigma_\Omega$ are constants which define the size of the modulation effect [1]. Details of $A_\Omega$ can also be found in Pope and Chen [59, 60] and for $A_{f,i}$ consult [39, 61, 62].

Using a procedure similar to the one used in deterministic case to derive Eq. (4.18), we obtain the MDF evolution equation for $\mathscr{F}_{fp}^E$. After integration over all particle or fluid properties results in the following equations for fluid and particle MDF evolution equations:

$$
\frac{\partial F_f^E}{\partial t} + \frac{\partial}{\partial x_i} \left[ v_{f,i} F_f^E \right] = -\frac{\partial}{\partial v_{f,i}} \left[ \left( A_{f,i} + \langle A_{p \to f,i} | \mathbf{Z}_f = \mathbf{z}_f \rangle \right) F_f^E \right]
$$
$$
- \frac{\partial}{\partial \theta} \left[ A_\Omega F_f^E + \langle A_{p \to \Omega} | \mathbf{Z}_f = \mathbf{z}_f \rangle F_f^E \right]
$$
$$
+ \frac{1}{2} \frac{\partial^2}{\partial \theta^2} \left[ (B_\Omega B_\Omega) F_f^E \right]
$$
$$
+ \frac{1}{2} \frac{\partial^2}{\partial v_{f,i} \partial v_{f,j}} \left[ \left( B_f B_f^T \right)_{ij} F_f^E \right] \tag{4.32}
$$

$$
\frac{\partial F_p^E}{\partial t} + \frac{\partial}{\partial x_i} \left[ v_{p,i} F_p^E \right] = -\frac{\partial}{\partial v_{p,i}} \left[ A_{p,i} F_p^E \right]
$$
$$
- \frac{\partial}{\partial v_{s,i}} \left[ \left( A_{s,i} + \langle A_{p \to s,i} | \mathbf{Z}_p = \mathbf{z}_p \rangle \right) F_f^E \right]
$$
$$
+ \frac{1}{2} \frac{\partial^2}{\partial v_{s,i} \partial v_{s,j}} \left[ \left( B_s B_s^T \right)_{ij} F_p^E \right]. \tag{4.33}
$$

### 4.3.4 Transport Equation of a General Property

To derive the macroscopic field equations for the fluid phase, it is first necessary to define the averaging operator for an arbitrary function of fluid element or particle properties $\mathscr{H}_k$. The mass weighted ensemble average $\langle \mathscr{H}_f \rangle$ is calculated by integrating the corresponding MDF as follows:

$$
\alpha_f(\mathbf{x}, t) \rho_f \langle \mathscr{H}_f \rangle (\mathbf{x}, t) = \int_{-\infty}^{+\infty} \mathscr{H}_f F_f^E (\mathbf{v}_f, \theta_f; \mathbf{x}, t) \, d\mathbf{v}_f d\theta_f \tag{4.34}
$$

For particulate phase, we have:

$$
\alpha_p(\mathbf{x}, t) \rho_p \langle \mathscr{H}_p \rangle (\mathbf{x}, t) = \int_{-\infty}^{+\infty} \mathscr{H}_p F_p^E (\mathbf{v}_p, \mathbf{v}_s, \delta_p; \mathbf{x}, t) \, d\mathbf{v}_p d\mathbf{v}_s d\delta_p. \tag{4.35}
$$

In addition, following averaging operations can be defined for an arbitrary function of the fluid and particle phase fluctuating properties, respectively,

$$
\alpha_f(\mathbf{x}, t)\, \rho_f \left\langle u'_{f,i_1} \dots u'_{f,i_n} \right\rangle (\mathbf{x}, t) = \int_{-\infty}^{+\infty} \prod_{k=1}^{n} v'_{f,i_k} F'^{E}_{f} \left(\mathbf{v}'_f, \theta'_f; \mathbf{x}, t\right) d\mathbf{v}'_f d\theta'_f,
$$

(4.36)

$$
\alpha_f(\mathbf{x}, t)\, \rho_f \left\langle u'_{p,i_1} \dots u'_{p,i_n} u'_{s,i_1} \dots u'_{s,i_m} \phi'^{l}_{p} \right\rangle (\mathbf{x}, t)
$$

$$
= \int_{-\infty}^{+\infty} \delta'^{l}_{f} \prod_{a=1}^{n} v'_{p,i_a} \prod_{b=1}^{m} v'_{s,i_b} F'^{E}_{p} \left(\mathbf{v}'_p, \mathbf{v}'_s, \delta'_p; \mathbf{x}, t\right) d\mathbf{v}'_p d\mathbf{v}'_s d\delta'_f.
$$

(4.37)

Having defined Eq. (4.34), we can now multiply Eq. (4.32) by $\mathcal{H}_f$ and then integrate as follows:

$$
\int_{-\infty}^{+\infty} \mathcal{H}_f \frac{\partial F^{E}_{f}}{\partial t} d\mathbf{v}_f d\theta_f + \int_{-\infty}^{+\infty} \mathcal{H}_f \frac{\partial}{\partial x_i} \left[ v_{f,i} F^{E}_{f} \right] d\mathbf{v}_f d\theta_f
$$

$$
= - \int_{-\infty}^{+\infty} \mathcal{H}_f \frac{\partial}{\partial v_{f,i}} \left[ \left( A_{f,i} + \left\langle A_{p \to f,i} | \mathbf{Z}_f = \mathbf{z}_f \right\rangle \right) F^{E}_{f} \right] d\mathbf{v}_f d\theta_f
$$

$$
- \int_{-\infty}^{+\infty} \mathcal{H}_f \frac{\partial}{\partial \theta} \left[ A_{\Omega} F^{E}_{f} + \left\langle A_{p \to \Omega} | \mathbf{Z}_f = \mathbf{z}_f \right\rangle F^{E}_{f} \right] d\mathbf{v}_f d\theta_f
$$

$$
+ \int_{-\infty}^{+\infty} \mathcal{H}_f \frac{1}{2} \frac{\partial^2}{\partial v_{f,i} \partial v_{f,j}} \left[ \left( B_f B_f^{T} \right)_{ij} F^{E}_{f} \right] d\mathbf{v}_f d\theta_f
$$

$$
+ \int_{-\infty}^{+\infty} \mathcal{H}_f \frac{1}{2} \frac{\partial^2}{\partial \theta^2} \left[ (B_{\Omega} B_{\Omega}) F^{E}_{f} \right] d\mathbf{v}_f d\theta_f.
$$

(4.38)

The first and second terms in Eq. 4.38 can be simplified by recognizing that $\mathcal{H}_f$ is not a function of time $t$ or position $\mathbf{x}$. Hence, for the first term, we have

$$
\int_{-\infty}^{+\infty} \mathcal{H}_f \frac{\partial F^{E}_{f}}{\partial t} d\mathbf{v}_f d\theta_f \implies \int_{-\infty}^{+\infty} \frac{\partial}{\partial t} \left[ \mathcal{H}_f F^{E}_{f} \right] d\mathbf{v}_f d\theta_f = \frac{\partial}{\partial t} \left[ \alpha_f \rho_f \left\langle \mathcal{H}_f \right\rangle \right],
$$

(4.39)

and similarly, for the second term,

$$\int_{-\infty}^{+\infty} \mathcal{H}_f \frac{\partial}{\partial x_i} \left[ v_{f,i} F_f^E \right] \mathrm{d}\mathbf{v}_f \mathrm{d}\theta_f \implies \int_{-\infty}^{+\infty} \frac{\partial}{\partial x_i} \left[ v_{f,i} \mathcal{H}_f F_f^E \right] \mathrm{d}\mathbf{v}_f \mathrm{d}\theta_f$$

$$= \frac{\partial}{\partial x_i} \left[ \alpha_f \rho_f \langle u_{f,i} \mathcal{H}_f \rangle \right]. \tag{4.40}$$

The third and fourth terms are more involved and require integration by parts. For example, the third term simplifies according to

$$\int_{-\infty}^{+\infty} \mathcal{H}_f \frac{\partial}{\partial v_{f,i}} \left[ \left( A_{f,i} + \langle A_{\mathrm{p}\to f,i} | \mathbf{Z}_f = \mathbf{z}_f \rangle \right) F_f \right] \mathrm{d}\mathbf{v}_f \mathrm{d}\theta_f$$

$$\implies = \left[ \mathcal{H}_f \left( A_{f,i} + \langle A_{\mathrm{p}\to f,i} | \mathbf{Z}_f = \mathbf{z}_f \rangle \right) \right]_{-\infty}^{+\infty}$$

$$- \int_{-\infty}^{+\infty} \left( A_{f,i} + \langle A_{\mathrm{p}\to f,i} | \mathbf{Z}_f = \mathbf{z}_f \rangle \right) \frac{\partial \mathcal{H}_f}{\partial v_{f,i}} F_f^E \mathrm{d}\mathbf{v}_f \mathrm{d}\theta_f$$

$$= -\alpha_f \rho_f \left\langle A_{f,i} \frac{\partial \mathcal{H}_f}{\partial u_{f,i}} \right\rangle - \int_{-\infty}^{+\infty} \langle A_{\mathrm{p}\to f,i} | \mathbf{Z}_f = \mathbf{z}_f \rangle \frac{\partial \mathcal{H}_f}{\partial v_{f,i}} F_f^E \mathrm{d}\mathbf{v}_f \mathrm{d}\theta_f,$$

$$\tag{4.41}$$

where integrals are assumed to converge to zero at $\pm\infty$. The interphase term is left in integral form at this stage. Similarly, for the fourth term, we have

$$\int_{-\infty}^{+\infty} \mathcal{H}_f \frac{\partial}{\partial \theta_f} \left[ \left( A_\Omega + \langle A_{\mathrm{p}\to\Omega} | \mathbf{Z}_f = \mathbf{z}_f \rangle \right) F_f \right] \mathrm{d}\mathbf{v}_f \mathrm{d}\theta_f$$

$$\implies = -\alpha_f \rho_f \left\langle A_\Omega \frac{\partial \mathcal{H}_f}{\partial \omega_f} \right\rangle - \int_{-\infty}^{+\infty} \langle A_{\mathrm{p}\to\Omega} | \mathbf{Z}_f = \mathbf{z}_f \rangle \frac{\partial \mathcal{H}_f}{\partial \theta_f} F_f^E \mathrm{d}\mathbf{v}_f \mathrm{d}\theta_f. \tag{4.42}$$

The fifth and sixth terms are diffusive terms and also require integration by parts. The fifth term can be expanded as

$$\int_{-\infty}^{+\infty} \mathcal{H}_f \frac{\partial^2}{\partial v_{f,i} \partial v_{f,j}} \left[ \left( B_f B_f^T \right)_{ij} F_f^E \right] \mathrm{d}\mathbf{v}_f \mathrm{d}\theta_f$$

$$\implies = \left[ \mathcal{H}_f \frac{\partial}{\partial v_{f,i}} \left[ \left( B_f B_f^T \right)_{ij} F_f^E \right] \right]_{-\infty}^{+\infty}$$

$$- \int\limits_{-\infty}^{+\infty} \frac{\partial \mathcal{H}_f}{\partial v_{f,i}} \frac{\partial}{\partial v_{f,i}} \left[ \left( B_f B_f^T \right)_{ij} F_f^E \right] d\mathbf{v}_f d\theta_f$$

$$= - \int\limits_{-\infty}^{+\infty} \frac{\partial \mathcal{H}_f}{\partial v_{f,i}} \frac{\partial}{\partial v_{f,i}} \left[ \left( B_f B_f^T \right)_{ij} F_f^E \right] d\mathbf{v}_f d\theta_f$$

$$= - \left[ \left( B_f B_f^T \right)_{ij} F_f^E \frac{\partial \mathcal{H}_f}{\partial u_{f,j}} \right]_{-\infty}^{+\infty}$$

$$+ \int\limits_{-\infty}^{+\infty} \left( B_f B_f^T \right)_{ij} \frac{\partial^2 \mathcal{H}_f}{\partial v_{f,i} v_{f,j}} F_f^E d\mathbf{v}_f d\theta_f$$

$$= \alpha_f \rho_f \left\langle \left( B_f B_f^T \right)_{ij} \frac{\partial^2 \mathcal{H}_f}{\partial v_{f,i} \partial v_{f,j}} \right\rangle, \qquad (4.43)$$

and similarly, for the sixth term,

$$\int\limits_{-\infty}^{+\infty} \mathcal{H}_f \frac{\partial^2}{\partial \theta_f^2} \left[ \left( B_\Omega B_\Omega^T \right) F_f^E \right] d\mathbf{v}_f d\theta_f \Longrightarrow \alpha_f \rho_f \left\langle \left( B_\Omega B_\Omega^T \right) \frac{\partial^2 \mathcal{H}_f}{\partial \omega_f^2} \right\rangle. \quad (4.44)$$

Combining terms, we have a macroscopic transport equation for the general function $\mathcal{H}_f$ of fluid properties

$$\frac{\partial}{\partial t} \left[ \alpha_f \rho_f \left\langle \mathcal{H}_f \right\rangle \right] + \frac{\partial}{\partial x_j} \left[ \alpha_f \rho_f \left\langle u_{f,j} \mathcal{H}_f \right\rangle \right]$$

$$= + \alpha_f \rho_f \left\langle A_{f,j} \frac{\partial \mathcal{H}_f}{\partial u_{f,j}} \right\rangle$$

$$+ \int\limits_{-\infty}^{+\infty} \left\langle A_{p \to f,j} | \mathbf{Z}_f = \mathbf{z}_f \right\rangle \frac{\partial \mathcal{H}_f}{\partial u_{f,j}} F_f^E$$

$$+ \frac{1}{2} \alpha_f \rho_f \left\langle \left( B_f B_f^T \right)_{ij} \frac{\partial^2 \mathcal{H}_f}{\partial u_{f,i} \partial u_{f,j}} \right\rangle$$

$$+ \alpha_f \rho_f \left\langle A_\Omega \frac{\partial \mathcal{H}_f}{\partial \omega_f} \right\rangle + \int\limits_{-\infty}^{+\infty} \left\langle A_{p \to \Omega} | \mathbf{Z}_f = \mathbf{z}_f \right\rangle \frac{\partial \mathcal{H}_f}{\partial \omega_f} F_f^E d\mathbf{v}_f d\theta_f$$

$$+ \frac{1}{2} \alpha_f \rho_f \left\langle \left( B_\Omega B_\Omega^T \right) \frac{\partial^2 \mathcal{H}_f}{\partial \theta_f^2} \right\rangle. \qquad (4.45)$$

The boundary assumption used to derive Eq. (4.45) is equivalent to assuming that probability of finding a point with infinite velocity in the phase space vector is zero meaning that the PDFs converge to zero when at least one component of the velocity goes to infinity and this should be included in the construction of the PDFs.

At this point, we exclude the turbulent frequency from the fluctuation equations as we will close the equation for turbulent frequency after writing the first moment equations, see Sect. 4.4.1.3. Writing Eq. (4.45) in terms of transformed coordinates

$$\mathbf{v}_f \rightarrow \mathbf{v}'_f \left(= \mathbf{v}_f - \langle \mathbf{u}_f \rangle (\mathbf{x}, t)\right),$$

makes the derivation of fluctuating field transport equations straightforward. However, we first need to define the derivatives in the transformed coordinates. By writing

$$\frac{\partial \mathscr{P}^E (\mathbf{v})}{\partial x_i} = \frac{\partial \mathscr{P}^E (\mathbf{v})}{\partial v'_j} \frac{\partial v'_j}{\partial x_i},$$

then replacing

$$\mathbf{v}' = \mathbf{v} - \langle \mathbf{u} \rangle (\mathbf{x}, t),$$

and noting

$$\mathscr{P}^E (\mathbf{v}) \, d\mathbf{v} = \mathscr{P}'^E (\mathbf{v}') \, d\mathbf{v}',$$

we have:

$$\frac{\partial \mathscr{P}^E (\mathbf{v})}{\partial x_i} = \frac{\partial \mathscr{P}'^E (\mathbf{v}')}{\partial x_i} - \frac{\partial \langle u_j \rangle}{\partial x_i} \frac{\partial \mathscr{P}'^E (\mathbf{v}')}{\partial v_j}, \tag{4.46}$$

where $\mathscr{P}'$ is the transformed PDF in the new transformed coordinate system. Now, we can transform Eq. (4.45) using Eq. (4.46) and definition (4.36) for an arbitrary fluctuating fluid property $\mathscr{H}'_f$ as follow:

$$\frac{D_f}{Dt} \left[\alpha_f \rho_f \left\langle \mathscr{H}'_f \right\rangle \right] + \frac{\partial}{\partial x_m} \left[\alpha_f \rho_f \left\langle u'_{f,m} \mathscr{H}'_f \right\rangle \right] = \alpha_f \rho_f \left\langle A_{f,m} \frac{\partial \mathscr{H}'_f}{\partial u'_{f,m}} \right\rangle$$

$$- \alpha_f \rho_f \frac{D_f \langle u_{f,m} \rangle}{Dt} \left\langle \frac{\partial \mathscr{H}'_f}{\partial u'_{f,m}} \right\rangle - \alpha_f \rho_f \frac{\partial \langle u_{f,n} \rangle}{\partial x_m} \left\langle \frac{\partial u'_{f,m} \mathscr{H}'_f}{\partial u'_{f,n}} \right\rangle$$

$$+ \frac{1}{2} \alpha_f \rho_f \left\langle \left(B_f B_f^T\right)_{mn} \frac{\partial^2 \mathscr{H}'_f}{\partial u'_{f,m} \partial u'_{f,n}} \right\rangle$$

$$+ \int_{-\infty}^{+\infty} \left(A_{f,m} + \langle A_{p \rightarrow f,m} | \mathbf{Z}_f = \mathbf{z}_f \rangle \right) \frac{\partial \mathscr{H}'_f}{\partial v'_{f,m}} F'^E_f d\mathbf{v}'_f, \tag{4.47}$$

where $D_k \langle \cdot \rangle / Dt = \partial \langle \cdot \rangle / \partial t + \langle u_{k,m} \rangle \partial \langle \cdot \rangle / \partial x_m$.

Using definitions (4.37) and (4.35), a similar procedure results in the following macroscopic equations for a general particle property $\mathscr{H}_p$ and a general particle fluctuating property $\mathscr{H}_p'$

$$\frac{\partial}{\partial t}\left[\alpha_p\rho_p\langle\mathscr{H}_p\rangle\right]+\frac{\partial}{\partial x_i}\left[\alpha_p\rho_p\langle\mathscr{H}_p\rangle\right]=\alpha_p\rho_p\left\langle A_{p,j}\frac{\partial\mathscr{H}_p}{\partial u_{p,j}}\right\rangle+\alpha_p\rho_p\left\langle A_{s,j}\frac{\partial\mathscr{H}_p}{\partial u_{s,j}}\right\rangle$$

$$+\frac{1}{2}\alpha_p\rho_p\left\langle\left(B_sB_s^T\right)_{ij}\frac{\partial^2\mathscr{H}_p}{\partial u_{s,i}\partial u_{s,j}}\right\rangle+\int_{-\infty}^{+\infty}\langle A_{p\to s,j}|\mathbf{Z}_p=\mathbf{z}_p\rangle\frac{\partial\mathscr{H}_p}{\partial v_{s,j}}F_p^E\,d\mathbf{v}_p d\mathbf{v}_s d\delta_p,$$

$$(4.48)$$

$$\frac{D_p}{Dt}\left[\alpha_p\rho_p\left\langle\mathscr{H}_p'\right\rangle\right]+\frac{\partial}{\partial x_m}\left[\alpha_p\rho_p\left\langle u_{p,m}'\mathscr{H}_p'\right\rangle\right]$$

$$=-\alpha_p\rho_p\left[\frac{D_p\langle u_{p,m}\rangle}{Dt}\left\langle\frac{\partial\mathscr{H}_p'}{\partial u_{p,m}'}\right\rangle+\frac{D_p\langle u_{s,m}\rangle}{Dt}\left\langle\frac{\partial\mathscr{H}_p'}{\partial u_{s,m}'}\right\rangle+\frac{D_p\langle\phi_p\rangle}{Dt}\left\langle\frac{\partial\mathscr{H}_p'}{\partial\phi_p'}\right\rangle\right]$$

$$+\alpha_p\rho_p\left[\left\langle A_{p,m}\frac{\partial\mathscr{H}_p'}{\partial u_{p,m}'}\right\rangle+\left\langle A_{s,m}\frac{\partial\mathscr{H}_p'}{\partial u_{s,m}'}\right\rangle\right]$$

$$-\alpha_p\rho_p\left[\frac{\partial\langle u_{p,n}\rangle}{\partial x_m}\left\langle\frac{\partial}{\partial u_{p,n}'}\left[u_{p,m}'\mathscr{H}_p'\right]\right\rangle+\frac{\partial\langle u_{s,n}\rangle}{\partial x_m}\left\langle\frac{\partial}{\partial u_{s,n}'}\left[u_{p,m}'\mathscr{H}_p'\right]\right\rangle\right.$$

$$\left.+\frac{\partial\langle\phi_p\rangle}{\partial x_m}\left\langle\frac{\partial}{\partial\phi_p'}\left[u_{p,m}'\mathscr{H}_p'\right]\right\rangle\right]$$

$$+\frac{1}{2}\alpha_p\rho_p\left\langle\left(B_sB_s^T\right)_{mn}\frac{\partial^2\mathscr{H}_p'}{\partial u_{s,m}'\partial u_{s,n}'}\right\rangle$$

$$+\int_{-\infty}^{+\infty}\langle A_{p\to s,m}|\mathbf{Z}_p=\mathbf{z}_p\rangle\frac{\partial\mathscr{H}_p'}{\partial u_{s,m}'}F_p'^E\,d\mathbf{v}_p'd\mathbf{v}_s'd\delta_p'. \tag{4.49}$$

In the next section, we will use these results to derive the mean field equations for up to third-order moments and discuss the results.

## 4.4 Fluid–Particle Field Equations

Using the results of the previous three sections, we are now in a position to easily derive the field equations for various field variables and their moments systematically. In this section, we will list the various equations using Eqs. (4.45) and (4.47) for fluid and (4.48) and (4.49) for particle phase directly.

### 4.4.1 Fluid Field Equations

#### 4.4.1.1 Fluid Continuity Equation

Setting $\mathscr{H}_f = 1$ in Eq. (4.45) results in the following equation for fluid continuity

$$\frac{\partial}{\partial t}\left(\alpha_f \rho_f\right) + \frac{\partial}{\partial x_i}\left[\alpha_f \rho_f \langle u_{f,i}\rangle\right] = 0 \qquad (4.50)$$

This is the usual continuity equation considering the volume fraction of fluid phase.

#### 4.4.1.2 Fluid Phase Momentum Equation

Setting $\mathscr{H}_f = u_{f,i}$ in Eq. (4.45) results in the following equation for fluid momentum using material derivative defined in previous section:

$$\alpha_f \rho_f \frac{D_f \langle u_{f,i}\rangle}{Dt} = -\frac{\partial}{\partial x_j}\left[\alpha_f \rho_f \langle u'_{f,i} u'_{f,j}\rangle\right] + \alpha_f \rho_f \langle A_{f,i}\rangle + S_{\text{p}\rightarrow f,i}^{\langle \mathbf{u}_f\rangle}, \qquad (4.51)$$

with $S_{\text{p}\rightarrow f,i}^{\langle \mathbf{u}_f\rangle} \equiv \int_{-\infty}^{+\infty} \langle A_{\text{p}\rightarrow f,i}|\mathbf{Z}_f = \mathbf{z}_f\rangle F_f^E d\mathbf{v}_f d\theta_f$. The first term in the RHS of Eq. (4.51) is the Reynolds stress tensor arising from the mean momentum flux due to the fluctuating velocity on the boundary of a control volume [61]. Second term is the fluid acceleration term which contains mean pressure gradient and viscous stresses, and the last term is the effect of the particle on the fluid velocity field. The second and third terms require modelling, and closures can be found in [63].

#### 4.4.1.3 Fluid Phase Turbulent Frequency

Similarly, using $\mathscr{H}_f = \omega_f$ in Eq. (4.45), we get the following equation for fluid turbulence frequency:

$$\alpha_f \rho_f \frac{D_f \langle \omega_f\rangle}{Dt} = -\frac{\partial}{\partial x_i}\left[\alpha_f \rho_f \langle \omega'_f u'_{f,i}\rangle\right] + \alpha_f \rho_f \langle A_\Omega\rangle + S_{\text{p}\rightarrow\Omega}^{\langle \omega_f\rangle}, \qquad (4.52)$$

where $S_{\text{p}\rightarrow\Omega}^{\langle \omega_f\rangle} \equiv \int_{-\infty}^{+\infty} \langle A_{\text{p}\rightarrow\Omega}|\mathbf{Z}_f = \mathbf{z}_f\rangle F_f^E d\mathbf{v}_f d\theta_f$. Eq. (4.52) contains the unclosed turbulent flux $\langle \omega'_f u'_{f,i}\rangle$ which requires modelling. Adopting the Langevin equation of Pope and Chen [60, 61], the following model equation is obtained [18] for $\langle \omega_f\rangle$ which is dimensionally equivalent to the model equation proposed by Hanjalic and Launder for $\langle \epsilon_f\rangle$ [64]:

$$\alpha_f \rho_f \frac{D_f \langle \omega_f \rangle}{Dt} = -\frac{\partial}{\partial x_i} \left[ \alpha_f \rho_f \frac{C_\omega}{\langle \omega_f \rangle} \langle u'_{f,i} u'_{f,i} \rangle \frac{\partial \langle \omega_f \rangle}{\partial x_j} \right]$$

$$+ \alpha_f \rho_f \left[ C_{\omega 1} \langle \omega_f \rangle \frac{P}{k_f} - \langle \omega_f \rangle^2 C_{\omega 2} \right] + S_{p \to \Omega}^{\langle \omega_f \rangle}, \qquad (4.53)$$

where $P$ is the production term. Using this closure, we close the turbulent frequency equation at this level, and thus, in writing the fluid fluctuation moments, we only consider the fluctuating term $\langle u'_{f,j} u'_{f,i} \rangle$ and exclude the turbulence frequency from the equations. This lets us to extract more information from the fluid field without making the equations more complicated. However, it would be interesting to keep this variable and write an explicit equation for $\langle \omega'_f u'_{f,i} \rangle$ which would then require closures for the third-order mixed moments of fluctuating velocity and turbulent frequency. We are not discussing this model in more detail here and refer the reader to [60, 61].

### 4.4.1.4 Fluid Phase Reynolds Stresses

Using Eq. (4.47) with $\mathcal{H}_f = u'_{f,i} u'_{f,j}$ and apply the material derivative results in:

$$\alpha_f \rho_f \frac{D_f}{Dt} \left[ \langle u'_{f,i} u'_{f,j} \rangle \right] = -\frac{\partial}{\partial x_k} \left[ \alpha_f \rho_f \langle u'_{f,i} u'_{f,j} u'_{f,k} \rangle \right]$$

$$- \alpha_f \rho_f \langle u'_{f,i} u'_{f,k} \rangle \frac{\partial \langle u_{f,j} \rangle}{\partial x_k} - \alpha_f \rho_f \langle u'_{f,j} u'_{f,k} \rangle \frac{\partial \langle u_{f,i} \rangle}{\partial x_k}$$

$$+ \alpha_f \rho_f \langle A_{f,i} u'_{f,j} + A_{f,j} u'_{f,i} \rangle + \alpha_f \rho_f \left\langle \left( B_f B_f^T \right)_{ij} \right\rangle + S_{p \to f,ij}^{\langle u'_{f,i} u'_{f,j} \rangle},$$

$$(4.54)$$

where

$$S_{p \to f,ij}^{\langle u'_{f,i} u'_{f,j} \rangle} \equiv \int_{-\infty}^{+\infty} \left[ v'_{f,i} \langle A_{p \to f,j} | \mathbf{Z}_f = \mathbf{z}_f \rangle + v'_{f,j} \langle A_{p \to f,i} | \mathbf{Z}_f = \mathbf{z}_f \rangle \right] F'^E_f dv'_f d\theta'_f.$$

$$(4.55)$$

In Eq. (4.54), the first term is the triple correlation of the fluctuating velocities, the second and third terms are the production terms, and the fourth term is the fluid acceleration term that requires modelling and contains the effects of both pressure gradient and fluctuating viscous terms [40]. The product of diffusion tensor contains the dissipation tensor, $\epsilon_{ij}$, should be modelled [40, 49, 65]. The triple correlation encountered in Eq. (4.54) is an unclosed term, and the logical starting point for a closure expression would be its transport equation. Derivation of the these transport correlations is the subject of Chap. 5. In the remaining part of this section, we explicitly derive Eq. (4.54) from the general transport equation for an arbitrary fluctuating property. Again, starting from Eq. (4.47), with $\mathcal{H}'_f = u'_{f,i} u'_{f,j}$, it is possible to write

$$\frac{D_f}{Dt}\left[\alpha_f\rho_f\left\langle u'_{f,i}u'_{f,j}\right\rangle\right]=\alpha_f\rho_f\frac{D_f}{Dt}\left[\left\langle u'_{f,i}u'_{f,j}\right\rangle\right]+\left\langle u'_{f,i}u'_{f,j}\right\rangle\frac{D_f}{Dt}[\alpha_f\rho_f]. \quad (4.56)$$

Using the definition of the material derivative, continuity equation for the fluid phase (Eq. (4.50)) can be written as

$$\frac{D_f}{Dt}\left(\alpha_f\rho_f\right)=-\alpha_f\rho_f\frac{\partial u_{f,k}}{\partial x_k}. \quad (4.57)$$

Replacing $\frac{D_f}{Dt}\left(\alpha_f\rho_f\right)$ in Eq. (4.56) using Eq. (4.57) gives

$$\frac{D_f}{Dt}\left[\alpha_f\rho_f\left\langle u'_{f,i}u'_{f,j}\right\rangle\right]=\alpha_f\rho_f\frac{D_f}{Dt}\left[\left\langle u'_{f,i}u'_{f,j}\right\rangle\right]-\alpha_f\rho_f\left\langle u'_{f,i}u'_{f,j}\right\rangle\frac{\partial u_{f,k}}{\partial x_k}. \quad (4.58)$$

The first term in the RHS of Eq. (4.58) is that appearing in the LHS of Eq. (4.54), and second term will be discussed later. Derivation of other terms is more involved, and derivations are as follows:

1. Drift term $\alpha_f\rho_f\langle A_{f,i}u'_{f,j}+A_{f,j}u'_{f,i}\rangle$: First note that the velocity components are independent variables; therefore, $\frac{\partial u'_{f,n}}{\partial u'_{f,m}}=\delta_{nm}$; using this fact, it is possible to write:

$$\alpha_f\rho_f\left\langle A_{f,m}\frac{\partial \mathcal{H}'_f}{\partial u'_{f,m}}\right\rangle=\alpha_f\rho_f\left\langle A_{f,m}\frac{\partial u'_{f,i}u'_{f,j}}{\partial u'_{f,m}}\right\rangle$$

$$=\alpha_f\rho_f\left\langle A_{f,m}u'_{f,j}\delta_{im}+A_{f,m}u'_{f,i}\delta_{jm}\right\rangle$$

$$=\alpha_f\rho_f\left\langle A_{f,i}u'_{f,j}+A_{f,j}u'_{f,i}\right\rangle.$$

2. Diffusion terms $\alpha_f\rho_f\left\langle\left(B_fB_f^T\right)_{ij}\right\rangle$:

$$\frac{1}{2}\alpha_f\rho_f\left\langle\left(B_fB_f^T\right)_{mn}\frac{\partial^2 \mathcal{H}'_f}{\partial u'_{f,m}\partial u'_{f,n}}\right\rangle$$

$$=\frac{1}{2}\alpha_f\rho_f\left\langle\left(B_fB_f^T\right)_{mn}\frac{\partial^2 u'_{f,i}u'_{f,j}}{\partial u'_{f,m}\partial u'_{f,n}}\right\rangle$$

$$=\frac{1}{2}\alpha_f\rho_f\left\langle\left(B_fB_f^T\right)_{mn}\frac{\partial}{\partial u'_{f,m}}(u'_{f,i}\delta_{jn}+u'_{f,j}\delta_{in})\right\rangle$$

$$=\frac{1}{2}\alpha_f\rho_f\left\langle\left(B_fB_f^T\right)_{mn}(\delta_{im}\delta_{jn}+\delta_{jm}\delta_{in})\right\rangle$$

$$=\frac{1}{2}\alpha_f\rho_f\left\langle\left(B_fB_f^T\right)_{ij}+\left(B_fB_f^T\right)_{ji}\right\rangle=\alpha_f\rho_f\left\langle\left(B_fB_f^T\right)_{ij}\right\rangle,$$

where the symmetry property of $\left(B_f B_f^T\right)$ is used to move to the last line.

3. Convection terms $-\alpha_f \rho_f \left\langle u'_{f,i} u'_{f,k}\right\rangle \frac{\partial \langle u_{f,j}\rangle}{\partial x_k} - \alpha_f \rho_f \left\langle u'_{f,j} u'_{f,k}\right\rangle \frac{\partial \langle u_{f,i}\rangle}{\partial x_k}$:

$$-\alpha_f \rho_f \frac{\partial \langle u_{f,n}\rangle}{\partial x_m} \left\langle \frac{\partial u'_{f,m} \mathscr{H}'_f}{\partial u'_{f,n}}\right\rangle = -\alpha_f \rho_f \frac{\partial \langle u_{f,n}\rangle}{\partial x_m} \left\langle \frac{\partial u'_{f,m} u'_{f,i} u'_{f,j}}{\partial u'_{f,n}}\right\rangle$$

$$= -\alpha_f \rho_f \frac{\partial \langle u_{f,n}\rangle}{\partial x_m} \left\langle u'_{f,m} u'_{f,i} \delta_{jn} + u'_{f,m} u'_{f,j} \delta_{in} + u'_{f,i} u'_{f,j} \delta_{mn}\right\rangle$$

$$= -\alpha_f \rho_f \left\langle u'_{f,i} u'_{f,k}\right\rangle \frac{\partial \langle u_{f,j}\rangle}{\partial x_k} - \alpha_f \rho_f \left\langle u'_{f,j} u'_{f,k}\right\rangle \frac{\partial \langle u_{f,i}\rangle}{\partial x_k}.$$

Changing the dummy index $m$ to $k$, the term $-\alpha_f \rho_f \left\langle u'_{f,i} u'_{f,j}\right\rangle \frac{\partial \langle u_{f,k}\rangle}{\partial x_k}$ cancels the corresponding term in the LHS (Eq. (4.58)).

4. Setting $\mathscr{H}'_f = u'_{f,i} u'_{f,j}$ in $\alpha_f \rho_f \frac{D_f \langle u_{f,m}\rangle}{Dt} \left\langle \frac{\partial \mathscr{H}'_f}{\partial u'_{f,m}}\right\rangle$ will produce first-order moments of fluctuating velocity which is zero by definition.

### 4.4.2 Particle Phase Mean Equations

#### 4.4.2.1 Particulate Phase Continuity Equation

Setting $\mathscr{H}_p = 1$ in Eq. (4.48) results in

$$\frac{\partial}{\partial t}\left[\alpha_p \rho_p\right] + \frac{\partial}{\partial x_i}\left[\alpha_p \rho_p \langle u_{p,i}\rangle\right] = 0. \tag{4.59}$$

This equation is simply the mass-balance equation for a particle phase without any interphase mass transfer or transfer due to coalescence and break-ups. Although we simplified the state vectors to reduce the complexity of the model, these effects can easily be included in the current framework by adding the appropriate variables to the state vector which will consequently appear on the RHS of Eq. (4.59) as integral source terms. Modelling approaches such as those proposed by Beck and Watkins [66–68] or Simonin [50] can be used to close these terms.

#### 4.4.2.2 Particulate Phase and Seen Momentum Equations

Setting $\mathscr{H}_p = u_{p,i}$ results in

$$\alpha_p \rho_p \frac{D_p \langle u_{p,i}\rangle}{Dt} = -\frac{\partial}{\partial x_j}\left[\alpha_p \rho_p \left\langle u'_{p,i} u'_{p,j}\right\rangle\right] + \alpha_p \rho_p \langle A_{p,i}\rangle. \tag{4.60}$$

The first term on the RHS of Eq. (4.60) is the particle velocity Reynolds stress tensor which can be interpreted, in a manner similar to the fluid Reynolds stress tensor, as being the momentum flux due to the fluctuating particle velocity. These terms require closures for which we obtain explicit equations. The second term is an acceleration term, examples include drag force, effects of mean pressure gradient and any external body forces such as gravity [50]. However, in this review, we only consider the drag force which is actually the only significant force due to the current modelling assumptions, see Sect. 4.1.2. Similarly, setting $\mathcal{H}_p = u_{s,i}$ for the seen velocity, we have

$$
\alpha_p \rho_p \frac{D_p \langle u_{s,i} \rangle}{Dt} = -\frac{\partial}{\partial x_j} \left[ \alpha_p \rho_p \langle u'_{s,i} u'_{p,j} \rangle \right] + \alpha_p \rho_p \langle A_{s,i} \rangle + S_{p \to s,i}^{\langle u_{s,i} \rangle},
\tag{4.61}
$$

where

$$
S_{p \to s,i}^{\langle u_{s,i} \rangle} \equiv \int_{-\infty}^{+\infty} \langle A_{p \to s,j} | \mathbf{Z}_p = \mathbf{z}_p \rangle F'_f^E dv'_p dv'_s d\delta'_p.
\tag{4.62}
$$

This latter equation accounts for the two-way coupling between the phases.

Equation (4.60) can explicitly be derived starting from Eq. (4.48). Replacing $\mathcal{H}_p = u_{p,i}$, the first term in the LHS of Eq. (4.48) can be written as

$$
\frac{\partial}{\partial t} [\alpha_p \rho_p \langle u_{p,i} \rangle] = \alpha_p \rho_p \frac{\partial}{\partial t} [\langle u_{p,i} \rangle] + \langle u_{p,i} \rangle \frac{\partial}{\partial t} [\alpha_p \rho_p].
\tag{4.63}
$$

Noting $u_{p,i} = \langle u_{p,i} \rangle + u'_{p,i}$ and that by definition, the first-order moments of the fluctuating component are zero and second term in Eq. (4.48) can be written by

$$
\begin{aligned}
\frac{\partial}{\partial x_j} \left[ \alpha_p \rho_p \langle u_{p,i} u_{p,j} \rangle \right] &= \frac{\partial}{\partial x_j} \left[ \alpha_p \rho_p \langle u'_{p,i} u'_{p,j} \rangle \right] + \frac{\partial}{\partial x_j} \left[ \alpha_p \rho_p \langle u_{p,i} \rangle \langle u_{p,j} \rangle \right] \\
&= \frac{\partial}{\partial x_j} \left[ \alpha_p \rho_p \langle u'_{p,i} u'_{p,j} \rangle \right] + \langle u_{p,i} \rangle \frac{\partial}{\partial x_j} \left[ \alpha_p \rho_p \langle u_{p,j} \rangle \right] \\
&\quad + \alpha_p \rho_p \langle u_{p,j} \rangle \frac{\partial}{\partial x_j} \left[ \langle u_{p,i} \rangle \right].
\end{aligned}
\tag{4.64}
$$

By multiplying the particle continuity equation [Eq. (4.59)] by $\langle u_{p,i} \rangle$, it is possible to write

$$
\langle u_{p,i} \rangle \frac{\partial}{\partial x_j} \left[ \alpha_p \rho_p \langle u_{p,j} \rangle \right] = -\langle u_{p,i} \rangle \frac{\partial}{\partial t} [\alpha_p \rho_p],
\tag{4.65}
$$

which cancels the second term in Eq. (4.63). Additional terms by definition reduce to the first term in the LHS and first term in the RHS of Eq. (4.60). By noting $u_{p,i}$ as an independent variable, the first term in the RHS of Eq. (4.48) can be simplified:

$$\alpha_p \rho_p \left\langle A_{p,j} \frac{\partial u_{p,i}}{\partial u_{p,j}} \right\rangle = \alpha_p \rho_p \left\langle A_{p,j} \delta_{ij} \right\rangle = \alpha_p \rho_p \left\langle A_{p,i} \right\rangle. \tag{4.66}$$

Other terms in the LHS of Eq. (4.48) are zero since the variables are independent, and this concludes the derivation of the Eq. (4.60).

### 4.4.2.3 Particle Mean Diameter

Setting $\mathscr{H}_p = \phi_p$ results in

$$\alpha_p \rho_p \frac{D_p \langle \phi_p \rangle}{Dt} = -\frac{\partial}{\partial x_i} \left[ \alpha_p \rho_p \left\langle \phi'_p u'_{p,i} \right\rangle \right]. \tag{4.67}$$

### 4.4.2.4 Particle Velocity Reynolds Stresses $\langle u'_{p,i} u'_{p,j} \rangle$

Setting $\mathscr{H}'_p = u'_{p,i} u'_{p,j}$ in Eq. (4.49) produces

$$\alpha_p \rho_p \frac{D_p}{Dt} \left[ \left\langle u'_{p,i} u'_{p,j} \right\rangle \right] = -\frac{\partial}{\partial x_k} \left[ \alpha_p \rho_p \left\langle u'_{p,i} u'_{p,j} u'_{p,k} \right\rangle \right] - \alpha_p \rho_p \left\langle u'_{p,i} u'_{p,k} \right\rangle \frac{\partial \langle u_{p,j} \rangle}{\partial x_k}$$
$$- \alpha_p \rho_p \left\langle u'_{p,j} u'_{p,k} \right\rangle \frac{\partial \langle u_{p,i} \rangle}{\partial x_k} + \alpha_p \rho_p \left\langle A_{p,i} u'_{p,j} + A_{p,j} u'_{p,i} \right\rangle, \tag{4.68}$$

for the particulate phase Reynolds stresses. First term on the RHS of Eq. (4.68) represents the transport of the stress by the particle fluctuating velocity which is diffusive in nature. This term can be written in terms of the second-order moments starting from the third-order moment transport equation and using a Boussinesq approximation [69]. Hence, the logical point to write closures for third-order moments is to start from their transport equation and this is the main reason for writing third-order moment transport equations for the current model. The second and third terms represent the production by the mean particle velocity gradient. The last term represents the effects of particle dragging on the fluid turbulence. The $A_p$ coefficients represent the drag forces and can be simply written as $\frac{1}{\tau_p} (u_{s,i} - u_{p,i})$, where $\tau_p$ is the timescale of the particles which depends on the diameter of the particle. If the diameters were constant, then we would have:

$$\alpha_p \rho_p \left\langle A_{p,i} u'_{p,j} + A_{p,j} u'_{p,i} \right\rangle$$
$$= \alpha_p \rho_p \left\langle \frac{1}{\tau_p} (u_{s,i} - u_{p,i}) u'_{p,j} + \frac{1}{\tau_p} (u_{u,j} - u_{u,j}) u'_{p,i} \right\rangle$$
$$= \frac{\alpha_p \rho_p}{\tau_p} \left[ \left\langle (u'_{s,i} + \langle u_{s,i} \rangle) u'_{p,j} \right\rangle - \left\langle (u'_{p,i} + \langle u_{p,i} \rangle) u'_{p,j} \right\rangle \right.$$

$$+ \left\langle \left( u'_{s,j} + \langle u_{s,j} \rangle \right) u'_{p,i} \right\rangle - \left\langle \left( u'_{p,j} + \langle u_{p,j} \rangle \right) u'_{p,i} \right\rangle \right]$$

$$= -\frac{2\alpha_p \rho_p}{\tau_p} \left\langle u'_{p,j} u'_{p,i} \right\rangle + \frac{\alpha_p \rho_p}{\tau_p} \left[ \left\langle u'_{s,i} u'_{p,j} \right\rangle + \left\langle u'_{s,j} u'_{p,i} \right\rangle \right]. \qquad (4.69)$$

This is exactly the equations derived by Simonin [57, 70] for the case of monodispersed particles. Thus, for the monodispersed systems, there is no need for any equation beyond Eqs. (4.68), (4.70) and (4.73). However, in the present model for a polydispersed particle population, here mixed moments of diameter and seen velocity, also diameter and particle velocity are needed which are derived in the next section. Writing the closures for these drag terms is not trivial and will be discussed in Chap. 5.

### 4.4.2.5  Particle Velocity/Seen Velocity Correlation $\left\langle u'_{s,i} u'_{p,j} \right\rangle$

By setting $\mathcal{H}'_p = u'_{s,i} u'_{p,j}$, we have:

$$\alpha_p \rho_p \frac{D_p}{Dt} \left[ \left\langle u'_{s,i} u'_{p,j} \right\rangle \right] = -\frac{\partial}{\partial x_k} \left[ \alpha_p \rho_p \left\langle u'_{s,i} u'_{p,j} u'_{p,k} \right\rangle \right]$$

$$- \alpha_p \rho_p \left\langle u'_{s,i} u'_{p,k} \right\rangle \frac{\partial \langle u_{p,j} \rangle}{\partial x_k} - \alpha_p \rho_p \left\langle u'_{p,j} u'_{p,k} \right\rangle \frac{\partial \langle u_{s,i} \rangle}{\partial x_k}$$

$$+ \alpha_p \rho_p \left\langle A_{s,i} u'_{p,j} + A_{p,j} u'_{s,i} \right\rangle + S_{p \to s, ij}^{\left\langle u'_{s,i} u'_{p,j} \right\rangle}, \qquad (4.70)$$

where

$$S_{p \to s, ij}^{\left\langle u'_{s,i} u'_{p,j} \right\rangle} \equiv \int\limits_{-\infty}^{+\infty} v'_{p,j} \left\langle A_{p \to s, j} | \mathbf{Z}_p = \mathbf{z}_p \right\rangle F'^{E}_p \, dv'_p dv'_s d\delta'_p. \qquad (4.71)$$

The first term in the RHS of Eq. (4.70) is the seen velocity–particle velocity correlation transported by the means of particle fluctuating velocity. Second and third terms are the production terms due to mean particle and seen velocity gradient. The fourth term contains different effects which are easier to interpret in the case of monodispersed particulate flows. Using analysis similar to Eq. (4.69), we can express the fourth term in Eq. (4.70) as a dissipation by the pressure–strain correlations $-1/\tau_f \left\langle u'_{s,i} u'_{p,j} \right\rangle$ where $\tau_f$ is the fluid Lagrangian integral timescale [70] and an interphase momentum transfer term $-\frac{1}{\tau_p} \left( \left\langle u'_{s,i} u'_{p,j} \right\rangle - \left\langle u'_{s,i} u'_{s,j} \right\rangle \right)$ which tends to bring the fluid–particle covariance, $\left\langle u'_{s,i} u'_{p,j} \right\rangle$, near to the fluid Reynolds stress tensor.

### 4.4.2.6 Particle Diameter/Particle Velocity Correlation $\left\langle \phi'_p u'_{p,i} \right\rangle$

By setting $\mathscr{H}'_p = \phi'_p u'_{p,i}$, we have

$$\alpha_p \rho_p \frac{D_p}{Dt} \left[ \left\langle \phi'_p u'_{p,i} \right\rangle \right] = -\frac{\partial}{\partial x_j} \left[ \alpha_p \rho_p \left\langle \phi'_p u'_{p,i} u'_{p,j} \right\rangle \right]$$
$$- \alpha_p \rho_p \left\langle \phi'_p u'_{p,j} \right\rangle \frac{\partial \left\langle u_{p,i} \right\rangle}{\partial x_j} - \alpha_p \rho_p \left\langle u'_{p,i} u'_{p,j} \right\rangle \frac{\partial \left\langle \phi_p \right\rangle}{\partial x_j}$$
$$+ \alpha_p \rho_p \left\langle A_{p,i} \phi'_p \right\rangle. \tag{4.72}$$

Again, this equation contains the triple correlation that can be interpreted as the transport of the particle diameter–particle velocity correlation, $\left\langle \phi'_p u'_{p,i} \right\rangle$, by the means of the particle fluctuating velocity. Second and third terms are the again production terms by the mean field gradients. The last term is the mixed effects of particle drag and the fluctuations in its diameter.

### 4.4.2.7 Seen Velocity Reynolds Stresses $\left\langle u'_{s,i} u'_{s,j} \right\rangle$

Equation (4.70) contains the term $\left\langle A_{p,j} u'_{s,i} \right\rangle$ which in turn will involve the seen velocity Reynolds stresses $\left\langle u'_{s,i} u'_{s,j} \right\rangle$, for which the transport equation is written as follows:

$$\alpha_p \rho_p \frac{D_p}{Dt} \left[ \left\langle u'_{s,i} u'_{s,j} \right\rangle \right] = -\frac{\partial}{\partial x_k} \left[ \alpha_p \rho_p \left\langle u'_{s,i} u'_{s,j} u'_{p,k} \right\rangle \right]$$
$$- \alpha_p \rho_p \left\langle u'_{s,i} u'_{s,k} \right\rangle \frac{\partial \left\langle u_{s,j} \right\rangle}{\partial x_k} - \alpha_p \rho_p \left\langle u'_{s,j} u'_{s,k} \right\rangle \frac{\partial \left\langle u_{s,i} \right\rangle}{\partial x_k}$$
$$+ \alpha_p \rho_p \left\langle A_{s,i} u'_{s,j} + A_{s,j} u'_{s,i} \right\rangle + \alpha_p \rho_p \left\langle (B_s B_s^T)_{ij} \right\rangle + S_{p \to s, ij}^{\left\langle u'_{s,i} u'_{s,j} \right\rangle}, \tag{4.73}$$

where

$$S_{p \to s, ij}^{\left\langle u'_{s,i} u'_{s,j} \right\rangle} \equiv \int\limits_{-\infty}^{+\infty} \left[ v'_{s,i} \left\langle A_{p \to s, j} | \mathbf{Z}_p = \mathbf{z}_p \right\rangle + v'_{s,j} \left\langle A_{p \to s, i} | \mathbf{Z}_p = \mathbf{z}_p \right\rangle \right] F'^E_p \, d\mathbf{v}'_p \, d\mathbf{v}'_s \, d\delta'_p. \tag{4.74}$$

This equation is similar to Eq. (4.54), and the terms can be interpreted similarly.

### 4.4.2.8 Particle Diameter/Seen Velocity Correlation $\left\langle \phi'_\mathrm{p} u'_{\mathrm{s},i} \right\rangle$

Equation (4.72) contains the term $\left\langle A_{\mathrm{p},j} \phi'_\mathrm{p} \right\rangle$ which in turn will involve the seen correlation $\left\langle \phi'_\mathrm{p} u'_{\mathrm{s},i} \right\rangle$, and the transport equation for $\left\langle \phi'_\mathrm{p} u'_{\mathrm{s},i} \right\rangle$ is:

$$
\alpha_\mathrm{p} \rho_\mathrm{p} \frac{D_\mathrm{p}}{Dt} \left[ \left\langle \phi'_\mathrm{p} u'_{\mathrm{s},i} \right\rangle \right] = -\frac{\partial}{\partial x_j} \left[ \alpha_\mathrm{p} \rho_\mathrm{p} \left\langle \phi'_\mathrm{p} u'_{\mathrm{s},i} u'_{\mathrm{p},j} \right\rangle \right]
$$

$$
- \alpha_\mathrm{p} \rho_\mathrm{p} \left\langle \phi'_\mathrm{p} u'_{\mathrm{p},j} \right\rangle \frac{\partial \left\langle u_{\mathrm{s},i} \right\rangle}{\partial x_j} - \alpha_\mathrm{p} \rho_\mathrm{p} \left\langle u'_{\mathrm{s},i} u'_{\mathrm{p},j} \right\rangle \frac{\partial \left\langle \phi_\mathrm{p} \right\rangle}{\partial x_j}
$$

$$
+ \alpha_\mathrm{p} \rho_\mathrm{p} \left\langle A_{\mathrm{s},i} \phi'_\mathrm{p} \right\rangle + S_{\mathrm{p} \rightarrow \mathrm{s},i}^{\left\langle \phi'_\mathrm{p} u'_{\mathrm{s},j} \right\rangle}, \tag{4.75}
$$

where

$$
S_{\mathrm{p} \rightarrow \mathrm{s},i}^{\left\langle \phi'_\mathrm{p} u'_{\mathrm{s},j} \right\rangle} \equiv \int_{-\infty}^{+\infty} \delta'_\mathrm{p} \left\langle A_{\mathrm{p} \rightarrow \mathrm{s},i} \middle| \mathbf{Z}_\mathrm{p} = \mathbf{z}_\mathrm{p} \right\rangle F'^{E}_\mathrm{p} \, \mathrm{dv}_\mathrm{p} \mathrm{dv}'_\mathrm{s} \mathrm{d}\delta'_\mathrm{p}. \tag{4.76}
$$

The first term is again transport of the particle diameter–seen velocity correlation, $\left\langle \phi'_\mathrm{p} u'_{\mathrm{s},i} \right\rangle$, by the particle velocity fluctuations, second and third terms are the production terms, and the fourth is the mixed effects of fluid seen acceleration and diameter fluctuations. It is also possible to write explicit analytical expressions involving different moments of diameter, seen velocity and particle velocity for the integral source terms in these equations by using the definition of conditional probability density and writing the expressions in terms of joint densities and marginals, see [1].

### 4.4.2.9 Particle Diameter Variance $\left\langle \phi'_\mathrm{p} \phi'_\mathrm{p} \right\rangle$

The particle diameter variance does not appear in any of the previous equations as an unclosed term. However, we require a transport equation for this term to describe the poly-dispersed nature of the particulate phase. This variance transport equation is obtained by setting $\mathcal{H}'_\mathrm{p} = \phi'_\mathrm{p} \phi'_\mathrm{p}$ as follows:

$$
\alpha_\mathrm{p} \rho_\mathrm{p} \frac{D_\mathrm{p}}{Dt} \left[ \left\langle \phi'_\mathrm{p} \phi'_\mathrm{p} \right\rangle \right] = -\frac{\partial}{\partial x_i} \left[ \alpha_\mathrm{p} \rho_\mathrm{p} \left\langle \phi'_\mathrm{p} \phi'_\mathrm{p} u'_{\mathrm{p},i} \right\rangle \right] - 2\alpha_\mathrm{p} \rho_\mathrm{p} \left\langle \phi'_\mathrm{p} u'_{\mathrm{p},i} \right\rangle \frac{\partial \left\langle \phi_\mathrm{p} \right\rangle}{\partial x_i}. \tag{4.77}
$$

## 4.5 Summary

In this chapter, the transport equations for a general mean and fluctuating property for the particle and fluid phases were derived from the Fokker–Planck equations. This is done by defining appropriate state vector and writing the corresponding SDEs.

Then, the results are applied to all possible permutation of the fluctuating and mean variables in the state vectors up to the second-order moments, which has resulted into 52 coupled PDEs. Expanding this framework for additional variables in the state vector, for instance, temperature and species mass fraction, is straightforward. It should be clear at this stage, which terms require closure in the current framework. These closures are the subject of the next chapter where third-order moment closure and non-integer moments are discussed in detail.

# References

1. Minier JP, Peirano E (2001) The pdf approach to turbulent polydispersed two-phase ows. Phys Rep 352:1–214
2. Pope SB (1994) On the relationship between stochastic Lagrangian models of turbulence and second-moment closures. Phys Fluids 6:973–985
3. Wells MR, Stock DE (1983) The effects of crossing trajectories on the dispersion of particles in a turbulent flow. J Fluid Mech 136:31–62
4. Csanady GT (1963) Turbulent diffusion of heavy particles in the atmosphere. J Atmos Sci 20:201–208
5. Boivin M, Simonin O, Squires KD (1998) Direct numerical simulation of turbulence modulation by particles in isotropic turbulence. J Fluid Mech 375:235–263
6. Stokes GG (1851) On the effect of the inertial friction of fluids on the motion of pendulums. Trans Camb Phil Soc 1
7. Gouesbet G, Berlemont A, Picart A (1984) Dispersion of discrete particles by continuous turbulent motion. extensive discussion of tchen's theory, using a two - parameter family of lagrangian correlation functions. Phys Fluids 27:827–837
8. Ranade VV (2002) Computational flow modeling for chemical reactor engineering. Academic Press, San Diego
9. Crowe C, Sommerfeld M, Tsuji Y (1998) Multiphase flows with droplets and particles. CRC Press, Boca Raton
10. Maxey MR, Riley JJ (1983) Equation of motion for small rigid sphere in non—uniform flow. Phys Fluids 4:883–889
11. Picart A, Berlemont A, Gouesbet G (1982) De linfulence du terme de basset sur la diepersion de particules discretes dans le cadre de la theorie de tchen. CR Acad Sci Paris Ser II:295–305
12. Rudinger G (1980) Handbook of powder technology. Elsevier Scientific Publishing Co, Amesterdam
13. Voir D, Michaelides E (1994) The effect of history term on the motion of rigid sphere in a viscous flow. Int J Multiph Flow 20:547
14. Auton TR (1983) The dynamics of bubbles, drops and particles in motion in liquids. Ph.D. thesis, University of Cambridge, Cambridge
15. Auton TR, Hunts JCR, Prud'homme M (1988) The force exerted on a body in inviscid unsteady non-uniform rotational flow. J Fluid Mech 197:241–257
16. Shrimpton JS, Yule AJ (1999) Characterisation of charged hydrocarbon sprays for application in combustion systems. Exp Fluids 26:460–469
17. Watts RG, Ferrer R (1987) The lateral force on a spinning sphere: aerodynamics of a curveball. Am J Phys 55:1–40
18. Scott SJ (2006) A pdf based method formodelling polysized particle laden turbulent flows without size class discretisation. PhD thesis, Imperial College London
19. Balachandar S (2009) A scaling analysis for pointparticle approaches to turbulent multiphase flows. Int J Multiph Flow 35:801–810

20. Ferrante A, Elghobashi S (2004) On the physical mechanism of drag reduction in a spatially developing turbulent boundary layer laden with microbubbles. J Fluid Mech 503:345–355
21. Wang L, Maxey M (1993) Settling velocity and concentration distribution of heavy particles in homogeneous isotropic turbulence. J Fluid Mech 256:27–68
22. Berrouk A, Laurence D, Riley J, Stock D (2007) Stochastic modeling of inertial particle dispersion by subgrid motion for les of high reynolds number pipe flow. J Turbul 8
23. Shotorban B, Balachandar S (2006) Particle concentration in homogeneous shear turbulence simulated via lagrangian and equilibrium Eulerian approaches. Phys Fluids 18
24. Minier JP, Peirano E, Chibbaro S (2004) Pdf model based on Langevin equation for polydispersed two-phase flows applied to a bluff-body gas-solid flow. Phys Fluids A 16:2419–2431
25. Chapman S, Cowling TG (1990) The mathematical theory Of non-uniform gases. Cambridge University Press, Cambridge
26. Gambosi TI (1994) Gas kinetic theory. Cambridge University Press, Cambridge
27. Bellomo N, Lo Schiavo M (1997) From the boltzmann equation to generalized kinetic models in applied sciences. Mathl Comput Model 26:43–76
28. Cercignani C (1975) Theory and application Of the boltzmann equation. Scottish Academic Press Ltd, New York
29. Cercignani C (1972) On the boltzmann equation for rigid spheres. Transp Theory Stat Phys 2:211–225
30. Gidaspow (1994) Multiphase flow and fluidization. Academic Press, San Diego
31. Frisch U (1991) Relation between the lattice boltzmann equation and the Navier–Stokes equations. Phys D 47:231–232
32. McKean HP (1969) A simple model of the derivation of fluid mechanics from the boltzmann equation. Bull Am Math Soc 75:1–10
33. Grad H (1964) Asymptotic theory of the boltzmann equation. Phys Fluids 6:147–181
34. Campbell CS (1990) Rapid granular flows. Ann Rev Fluid Meeh 22:57–92
35. Forterre Y, Pouliquen O (2008) Flows of dense granular media. Ann Rev Fluid Mech 40:1–24
36. Goldhirsch I (2003) Rapid granular flows. Ann Rev Fluid Mech 35:267–293
37. Neri A, Gidaspow D (2000) Riser hydrodynamics: simulation using kinetic theory. AIChE J 46:52–67
38. Subramaniam S (2000) Statistical representation of a spray as a point process. Phys Fluids 12:2413–2431
39. Peirano E, Chibbaro S, Pozorski J, Minier JP (2006) Mean-field/pdf numerical approach for polydispersed turbulent two-phase flows. Prog EnergyCombust Sci 32:315–371
40. Pope S (1994) On the relationship between stochastic lagrangian models of turbulence and second-moment closures. Phys Fluids 6:973–985
41. Csanady G (1963) Turbulent diffusion of heavy particles in the atmosphere. J Atmos Sci 20:201–208
42. Williams FA (1958) Spray combustion and atomization. Phys Fluids 1:541–545
43. Archambault MR, Edwards CF (2000) Computation of spray dynamics by direct solution of moment transport equations-inclusion of nonlinear momentum exchange. In: Eighth international conference on liquid atomization and spray systems
44. Archambault MR, Edwards CF, McCormack RW (2003) Computation of spray dynamics by moment transport equations i: theory and development. Atomization Sprays 13:63–87
45. Archambault MR, Edwards CF, McCormack RW (2003) Computation of spray dynamics by moment transport equations ii: application to quasi-one dimensional spray. Atomization Sprays 13:89–115
46. Domelevo K (2001) The kinetic sectional approach for noncolliding evaporating sprays. Atomization Sprays 11:291–303
47. Tambour Y (1980) A sectional model for evaporation and combustion of sprays of liquid fuels. Israel J Tech 18:47–56
48. Haken H (1989) Synergetics: an overview. Rep Prog Phys 52:515–533
49. Pope SB (1994) Lagrangian pdf methods for turbulent flows. Ann Rev Fluid Mech 26:23–63

50. Simonin O (2000) Statistical and continuum modelling of turbulent reactive particulate flows. part 1: theoretical derivation of dispersed phase Eulerian modelling from probability density function kinetic equation. In: Lecure series, Von-Karman Institute for Fluid Dynamics
51. Gosman AD, Ioannides E (1983) Aspects of computer simulation of liquid-fueled combustors. J Energy 7:482–490
52. Berlemont A, Desjonqueres P, Guesbet G (1990) Particle lagrangian simulation in turbulent flows. Int J Multiph Flow 16:19–34
53. Ormancey A, Martinon A (1984) Prediction of particle dispersion in turbulent flows. Phys-Chem Hydrodyn 5:229–244
54. Simonin O (2000) Statistical and continuum modelling of turbulent reactive particulate flows. part 2: application of a two-phase second-moment transport model for prediction of turbulent gas-particle flows. In: Lecure series, Von-Karman Institute for Fluid Dynamics
55. Perkins R, Ghosh S, Phillips J (1991) Interaction of particles and coherent structures in a plane turbulent air jet. Adv Turbul 3:93–100
56. Simonin O, Deutsch E, Minier JP (1993) Eulerian prediction of the fluid/particle correlated motion in turbulent two-phase flows. App Sci Res 51:275–283
57. Simonin O, Deutsch E, Bovin M (1993) Large eddy simulation and second-moment closure model of particle fluctuating motion in two-phase turbulent shear flows. Turbulent Shear Flows 9
58. Pope SB (1985) Pdf methods for turbulent reactive flows. Prog EnergyCombust Sci 11:119–192
59. Pope S (1991) Application of the velocity-dissipation probability density function model in inhomogeneous turbulent flows. Phys Fluids A 3(8):1947–1957
60. Pope SB, Chen YL (1990) The velocity-dissipation probability density function model for turbulent flows. Phys Fluids A 2(8):1437–1449
61. Pope S (2001) Turbulent flows. Cambridge University Press, Cambridge
62. Minier JP, Pozorski J (1997) Derivation of a pdf model for turbulent flows based on principles from statistical physics. Phys Fluids 9(6):1748–1753
63. Peirano E, Leckner B (1998) Fundamentals of turbulent gas-solid flows applied to circulating fluidized bed combustion. Prog Energy Combust Sci 24:259–296
64. Hanjalic K, Launder BE (1972) A reynolds stress model of turbulence and its application to thin shear flows. J Fluid Mech 52:609–638
65. Wouters HA, Peeters TWJ, Roekaerts D (1996) On the existence of a generalized langevin model representation for second-moment closures. Phys Fluids 8:1702–1704
66. Beck JC, Watkins AP (1999) Spray modelling using the moments of the droplet size distribution. In: ILASS-Europe
67. Beck JC, Watkins AP (2000) Modelling polydispersed sprays without discretisation into droplet size classes. In: ILASS Pasadena
68. Beck JC, Watkins AP (2003) The droplet number moments approach to spray modelling: the development of heat and mass transfer sub-models. Int J Heat Fluid Flow 24:242–259
69. Wang Q, Squires KD, Simonin O (1998) Large eddy simulation of turbulent gas-solid flows in a vertical channel and evaluation of second-order models. Int J Heat Fluid Flow 19:505–511
70. Simonin O, Deutsch E, Boivin M (1995) Comparison of large-eddy simulation and second-moment closure of particle fluctuating motion in two-phase turbulent shear flows. In: Turbulence and shear flows, vol 9. Springer, Berlin, pp 85–115

# Chapter 5
# Closure Problem

In Chap. 4, we explained that we are interested in the EE field equation up to second-order moments. These equations contain third-order moments, and the starting point for writing the closure for these moments would be their transport equations. In addition, the non-integer closure problem which is the result of the variable particle diameter will be discussed in this chapter.

## 5.1 Fluid Phase Triple Correlation

The fluid phase Reynolds stress equation (Eq. 4.54) contains the unclosed triple correlation of fluctuating fluid velocities $\left\langle u'_{f,i} u'_{f,j} u'_{f,k} \right\rangle$. Closure for this term is not explicitly provided in this book but the transport equation for this tensor provides the logical start point for obtaining an expression. Hence for future development of this framework, the third-order transport equations for both phases are provided in this chapter. In the single phase literature, several contracted expressions for $\left\langle u'_{f,i} u'_{f,j} u'_{f,k} \right\rangle$ have been proposed (e.g. [1]) and would be appropriate in the context of this work.

The fluid phase triple correlation fequation requires a more in depth manipulation of Eq. 4.47. Taking $\mathscr{H}'_f = u_{f,i} u_{f,j} u_{f,k}$ the first term of Eq. 4.47 becomes

$$\frac{\mathrm{D}_f}{\mathrm{D}t} \left[ \alpha_f \rho_f \left\langle \mathscr{H}'_f \right\rangle \right] \implies \frac{\mathrm{D}_f}{\mathrm{D}t} \left[ \alpha_f \rho_f \left\langle u'_{f,i} u'_{f,j} u'_{f,k} \right\rangle \right], \tag{5.1}$$

and similarly the second term becomes

$$\frac{\partial}{\partial x_m} \left[ \alpha_f \rho_f \left\langle u'_{f,m} \mathscr{H}'_f \right\rangle \right] \implies \frac{\partial}{\partial x_m} \left[ \alpha_f \rho_f \left\langle u'_{f,i} u'_{f,j} u'_{f,k} u'_{f,m} \right\rangle \right]. \tag{5.2}$$

The third term is more involved and can be expanded making use of the Dirac delta function $\delta_{ij}$ as follows:

$$\alpha_f \rho_f \left\langle A_{f,m} \frac{\partial \mathscr{H}'_f}{\partial u'_{f,m}} \right\rangle \Longrightarrow \alpha_f \rho_f \left\langle A_{f,m} \frac{\partial}{\partial u'_{f,m}} \left[ u'_{f,i} u'_{f,j} u'_{f,k} \right] \right\rangle$$

$$= \alpha_f \rho_f \left\langle A_{f,m} \left( u'_{f,i} u'_{f,j} \delta_{mk} + u'_{f,i} u'_{f,k} \delta_{mj} + u'_{f,j} u'_{f,k} \delta_{mi} \right) \right\rangle$$

$$= \alpha_f \rho_f \left\langle A_{f,i} u'_{f,j} u'_{f,k} + A_{f,j} u'_{f,i} u'_{f,k} + A_{f,k} u'_{f,i} u'_{f,j} \right\rangle. \tag{5.3}$$

The fourth term is shown to equal zero after expansion

$$\frac{1}{2}\alpha_f \rho_f \left\langle (B_f B_f^T)_{mn} \frac{\partial^2 \mathscr{H}'_f}{\partial u'_{f,m} \partial u'_{f,n}} \right\rangle \Longrightarrow \frac{1}{2}\alpha_f \rho_f \left\langle (B_f B_f^T)_{mn} \frac{\partial^2}{\partial v'_{f,m} \partial v'_{f,n}} \left[ u'_{f,i} u'_{f,j} u'_{f,k} \right] \right\rangle$$

$$= \frac{1}{2}\alpha_f \rho_f (B_f B_f^T)_{mn} \left\langle u'_{f,i} (\delta_{mj}\delta_{nk} + \delta_{mk}\delta_{nj}) \right.$$

$$\left. + u'_{f,j} (\delta_{mi}\delta_{nk} + \delta_{mk}\delta_{ni}) + u'_{f,k} (\delta_{mi}\delta_{nj} + \delta_{mj}\delta_{ni}) \right\rangle$$

$$= \frac{1}{2}\alpha_f \rho_f \left[ \left\langle u'_{f,i} \right\rangle (B_f B_f^T)_{jk} + \left\langle u'_{f,j} \right\rangle (B_f B_f^T)_{ik} \right.$$

$$\left. + \left\langle u'_{f,k} \right\rangle (B_f B_f^T)_{ij} \right]$$

$$= 0. \tag{5.4}$$

Expansion of the fifth term results in

$$\alpha_f \rho_f \frac{D_f \langle u_{f,m} \rangle}{Dt} \left\langle \frac{\partial \mathscr{H}'_f}{\partial u'_{f,m}} \right\rangle \Longrightarrow \alpha_f \rho_f \frac{D_f \langle u_{f,m} \rangle}{Dt} \left\langle \frac{\partial}{\partial u'_{f,m}} \left[ u'_{f,i} u'_{f,j} u'_{f,k} \right] \right\rangle$$

$$= \alpha_f \rho_f \frac{D_f \langle u_{f,m} \rangle}{Dt} \left\langle u'_{f,i} u'_{f,j} \delta_{mk} + u'_{f,i} u'_{f,k} \delta_{mj} + u'_{f,j} u'_{f,k} \delta_{mi} \right\rangle$$

$$= \alpha_f \rho_f \left[ \frac{D_f \langle u_{f,i} \rangle}{Dt} \left\langle u'_{f,j} u'_{f,k} \right\rangle + \frac{D_f \langle u_{f,j} \rangle}{Dt} \left\langle u'_{f,i} u'_{f,k} \right\rangle \right.$$

$$\left. + \frac{D_f \langle u_{f,k} \rangle}{Dt} \left\langle u'_{f,i} u'_{f,j} \right\rangle \right], \tag{5.5}$$

and similarly for the sixth we have

$$\alpha_f \rho_f \frac{\partial \langle u_{f,n} \rangle}{\partial x_m} \left\langle \frac{\partial}{\partial u'_{f,n}} \left[ u'_{f,m} \mathscr{H}'_f \right] \right\rangle \Longrightarrow \alpha_f \rho_f \frac{\partial \langle u_{f,n} \rangle}{\partial x_m} \left\langle \frac{\partial}{\partial u'_{f,n}} \left[ u'_{f,i} u'_{f,j} u'_{f,k} u'_{f,m} \right] \right\rangle$$

$$= \alpha_f \rho_f \frac{\partial \langle u_{f,n} \rangle}{\partial x_m} \left\langle u'_{f,i} u'_{f,j} u'_{f,k} \delta_{nm} + u'_{f,i} u'_{f,j} u'_{f,m} \delta_{nk} \right.$$

$$\left. + u'_{f,i} u'_{f,k} u'_{f,m} \delta_{nj} + u'_{f,j} u'_{f,k} u'_{f,m} \delta_{ni} \right\rangle$$

$$= \alpha_f \rho_f \left[ \left\langle u'_{f,i} u'_{f,k} u'_{f,m} \right\rangle \frac{\partial \langle u_{f,i} \rangle}{\partial x_m} + \left\langle u'_{f,i} u'_{f,k} u'_{f,m} \right\rangle \frac{\partial \langle u_{f,j} \rangle}{\partial x_m} \right.$$

$$\left. + \left\langle u'_{f,i} u'_{f,j} u'_{f,m} \right\rangle \frac{\partial \langle u_{f,k} \rangle}{\partial x_m} \right]. \tag{5.6}$$

Combining the above terms leads to the final form of the transport equation for the fluid velocity triple correlations

$$\alpha_f \rho_f \frac{D_f}{Dt} \left[ \left\langle u'_{f,i} u'_{f,j} u'_{f,k} \right\rangle \right] = -\frac{\partial}{\partial x_m} \left[ \alpha_f \rho_f \left\langle u'_{f,i} u'_{f,j} u'_{f,k} u'_{f,m} \right\rangle \right]$$

$$+ \alpha_f \rho_f \left\langle A_{f,i} u'_{f,j} u'_{f,k} + A_{f,j} u'_{f,i} u'_{f,k} + A_{f,k} u'_{f,i} u'_{f,j} \right\rangle$$

$$- \alpha_f \rho_f \left[ \frac{D_f \langle u_{f,i} \rangle}{Dt} \left\langle u'_{f,j} u'_{f,k} \right\rangle + \frac{D_f \langle u_{f,j} \rangle}{Dt} \left\langle u'_{f,i} u'_{f,k} \right\rangle \right.$$

$$\left. + \frac{D_f \langle u_{f,k} \rangle}{Dt} \left\langle u'_{f,i} u'_{f,j} \right\rangle \right]$$

$$- \alpha_f \rho_f \left[ \left\langle u'_{f,i} u'_{f,k} u'_{f,m} \right\rangle \frac{\partial \langle u_{f,i} \rangle}{\partial x_m} + \left\langle u'_{f,i} u'_{f,k} u'_{f,m} \right\rangle \frac{\partial \langle u_{f,j} \rangle}{\partial x_m} \right.$$

$$\left. + \left\langle u'_{f,i} u'_{f,j} u'_{f,m} \right\rangle \frac{\partial \langle u_{f,k} \rangle}{\partial x_m} \right] + S_{p \to f,ijk}^{\left\langle u'_{f,i} u'_{f,j} u'_{f,k} \right\rangle}, \tag{5.7}$$

where

$$S_{p \to f,ijk}^{\left\langle u'_{f,i} u'_{f,j} u'_{f,k} \right\rangle} \equiv \int_{-\infty}^{+\infty} \left[ v'_{f,i} v'_{f,j} \left\langle A_{p \to f,k} | \mathbf{Z}_f = \mathbf{z}_f \right\rangle + v'_{f,i} v'_{f,k} \left\langle A_{p \to f,j} | \mathbf{Z}_f = \mathbf{z}_f \right\rangle \right.$$

$$\left. + v'_{f,j} v'_{f,k} \left\langle A_{p \to f,i} | \mathbf{Z}_f = \mathbf{z}_f \right\rangle \right] F'^E_f \mathrm{d} \mathbf{v}'_f \mathrm{d} \theta'_f. \tag{5.8}$$

## 5.2 Particle Phase Triple Correlations

### 5.2.1 Particle Velocity Triple Correlation $\left\langle u'_{p,i} u'_{p,j} u'_{p,k} \right\rangle$

The particle velocity triple correlation $\left\langle u'_{p,i} u'_{p,j} u'_{p,k} \right\rangle$ transport equation is obtained by setting $\mathscr{H}'_p = u'_{p,i} u'_{p,j} u'_{p,k}$ as follows

$$\alpha_p \rho_p \frac{D_p}{Dt} \left[ \left\langle u'_{p,i} u'_{p,j} u'_{p,k} \right\rangle \right] = -\frac{\partial}{\partial x_m} \left[ \alpha_p \rho_p \left\langle u'_{p,i} u'_{p,j} u'_{p,k} u'_{p,m} \right\rangle \right]$$

$$+ \alpha_p \rho_p \left\langle A_{p,i} u'_{p,j} u'_{p,k} + A_{p,j} u'_{p,i} u'_{p,k} + A_{p,k} u'_{p,i} u'_{p,j} \right\rangle$$

$$- \alpha_p \rho_p \left[ \frac{D_p \langle u_{p,i} \rangle}{Dt} \left\langle u'_{p,j} u'_{p,k} \right\rangle + \frac{D_p \langle u_{p,j} \rangle}{Dt} \left\langle u'_{p,i} u'_{p,k} \right\rangle \right.$$

$$\left. + \frac{D_p \langle u_{p,k} \rangle}{Dt} \left\langle u'_{p,i} u'_{p,j} \right\rangle \right]$$

$$- \alpha_p \rho_p \left[ \left\langle u'_{p,i} u'_{p,k} u'_{p,m} \right\rangle \frac{\partial \langle u_{p,i} \rangle}{\partial x_m} + \left\langle u'_{p,i} u'_{p,k} u'_{p,m} \right\rangle \frac{\partial \langle u_{p,j} \rangle}{\partial x_m} \right.$$

$$\left. + \left\langle u'_{p,i} u'_{p,j} u'_{p,m} \right\rangle \frac{\partial \langle u_{p,k} \rangle}{\partial x_m} \right]. \tag{5.9}$$

## 5.2.2  Mixed Triple Correlation $\left\langle u'_{s,i} u'_{p,j} u'_{p,k} \right\rangle$

The mixed velocity triple correlation $\left\langle u'_{s,i} u'_{p,j} u'_{p,k} \right\rangle$ transport equation is obtained by setting $\mathcal{H}'_p = u'_{p,i} u'_{p,j} u'_{s,k}$ as follows

$$\alpha_p \rho_p \frac{D_p}{Dt} \left[ \left\langle u'_{s,i}, u'_{p,j}, u'_{p,k} \right\rangle \right] = -\frac{\partial}{\partial x_m} \left[ \alpha_p \rho_p \left\langle u'_{s,i} u'_{p,j} u'_{p,k} u'_{p,m} \right\rangle \right]$$

$$+ \alpha_p \rho_p \left\langle A_{s,i} u'_{p,j} u'_{p,k} + A_{p,j} u'_{s,i} u'_{p,k} + A_{p,k} u'_{s,i} u'_{p,j} \right\rangle$$

$$- \alpha_p \rho_p \left[ \frac{D_p \langle u_{s,i} \rangle}{Dt} \left\langle u'_{p,j} u'_{p,k} \right\rangle + \frac{D_p \langle u_{p,j} \rangle}{Dt} \left\langle u'_{s,i} u'_{p,k} \right\rangle \right.$$

$$\left. + \frac{D_p \langle u_{p,k} \rangle}{Dt} \left\langle u'_{s,i} u'_{p,j} \right\rangle \right]$$

$$- \alpha_p \rho_p \left[ \left\langle u'_{p,j} u'_{p,k} u'_{p,m} \right\rangle \frac{\partial \langle u_{s,i} \rangle}{\partial x_m} + \left\langle u'_{s,i} u'_{p,k} u'_{p,m} \right\rangle \frac{\partial \langle u_{p,j} \rangle}{\partial x_m} \right.$$

$$\left. + \left\langle u'_{s,i} u'_{p,j} u'_{p,m} \right\rangle \frac{\partial \langle u_{p,k} \rangle}{\partial x_m} \right] + S_{p \to s, ijk}^{\langle u'_{s,i} u'_{p,j} u'_{p,k} \rangle}$$

$$\tag{5.10}$$

where

$$S_{p \to s, ijk}^{\langle u'_{s,i} u'_{p,j} u'_{p,k} \rangle} \equiv \int\limits_{-\infty}^{+\infty} v'_{p,j} v'_{p,k} \left\langle A_{p \to s,i} | \mathbf{Z}_s = \mathbf{z}_s \right\rangle F'^{E}_s \, d\mathbf{v}'_p d\mathbf{v}'_s d\delta'_p. \tag{5.11}$$

### 5.2.3 Mixed Triple Correlation $\left\langle \phi'_p u'_{p,i} u'_{p,j} \right\rangle$

The mixed velocity-diameter triple correlation $\left\langle \phi'_p u'_{p,i} u'_{p,j} \right\rangle$ transport equation is obtained by setting $\mathcal{H}'_p = \phi'_p u'_{p,i} u'_{p,j}$ as follows

$$
\alpha_p \rho_p \frac{D_p}{Dt} \left[ \left\langle \phi'_p u'_{p,i} u'_{p,j} \right\rangle \right] = -\frac{\partial}{\partial x_k} \left[ \alpha_p \rho_p \left\langle \phi'_p u'_{p,i} u'_{p,j} u'_{p,k} \right\rangle \right]
$$
$$
+ \alpha_p \rho_p \left\langle A_{p,i} \phi'_p u'_{p,j} + A_{p,j} \phi'_p u'_{p,i} \right\rangle
$$
$$
- \alpha_p \rho_p \left[ \frac{D_p \langle u_{p,i} \rangle}{Dt} \left\langle \phi'_p u'_{p,j} \right\rangle + \frac{D_p \langle u_{p,j} \rangle}{Dt} \left\langle \phi'_p u'_{p,i} \right\rangle \right.
$$
$$
\left. + \frac{D_p \langle \phi_p \rangle}{Dt} \left\langle u'_{p,i} u'_{p,j} \right\rangle \right]
$$
$$
- \alpha_p \rho_p \left[ \left\langle \phi'_p u'_{p,i} u'_{p,k} \right\rangle \frac{\partial \langle u_{p,i} \rangle}{\partial x_k} + \left\langle \phi'_p u'_{p,j} u'_{p,k} \right\rangle \frac{\partial \langle u_{p,j} \rangle}{\partial x_k} \right.
$$
$$
\left. + \left\langle u'_{p,i} u'_{p,j} u'_{p,k} \right\rangle \frac{\partial \langle \phi_p \rangle}{\partial x_k} \right]. \tag{5.12}
$$

### 5.2.4 Mixed Triple Correlation $\left\langle u'_{s,i} u'_{s,j} u'_{p,k} \right\rangle$

The particle velocity seen triple correlation $\left\langle u'_{s,i} u'_{s,j} u'_{p,k} \right\rangle$ transport equation is obtained by setting $\mathcal{H}'_p = u'_{s,i} u'_{s,j} u'_{p,k}$ as follows

$$
\alpha_p \rho_p \frac{D_p}{Dt} \left[ \left\langle u'_{s,i} u'_{s,j} u'_{p,k} \right\rangle \right] = -\frac{\partial}{\partial x_m} \left[ \alpha_p \rho_p \left\langle u'_{s,i} u'_{s,j} u'_{p,k} u'_{p,m} \right\rangle \right]
$$
$$
+ \alpha_p \rho_p \left\langle A_{s,i} u'_{s,j} u'_{p,k} + A_{s,j} u'_{s,i} u'_{p,k} + A_{p,k} u'_{s,i} u'_{s,j} \right\rangle
$$
$$
- \alpha_p \rho_p \left[ \frac{D_p \langle u_{s,i} \rangle}{Dt} \left\langle u'_{s,j} u'_{p,k} \right\rangle + \frac{D_p \langle u_{s,j} \rangle}{Dt} \left\langle u'_{s,i} u'_{p,k} \right\rangle \right.
$$
$$
\left. + \frac{D_p \langle u_{p,k} \rangle}{Dt} \left\langle u'_{s,i} u'_{s,j} \right\rangle \right]
$$
$$
- \alpha_p \rho_p \left[ \left\langle u'_{s,i} u'_{p,k} u'_{p,m} \right\rangle \frac{\partial \langle u_{s,i} \rangle}{\partial x_m} + \left\langle u'_{s,i} u'_{p,k} u'_{p,m} \right\rangle \frac{\partial \langle u_{s,j} \rangle}{\partial x_m} \right.
$$
$$
\left. + \left\langle u'_{s,i} u'_{s,j} u'_{p,m} \right\rangle \frac{\partial \langle u_{s,k} \rangle}{\partial x_m} \right] + S^{\langle u'_{s,i} u'_{s,j} u'_{p,k} \rangle}_{p \rightarrow s, ijk}. \tag{5.13}
$$

where

$$S^{\langle u'_{s,i}u'_{s,j}u'_{p,k}\rangle}_{p\to s,ijk} \equiv \int\limits_{-\infty}^{+\infty} \Big[ v'_{s,i}v'_{p,k}\langle A_{p\to s,j}|\mathbf{Z}_s = \mathbf{z}_s\rangle$$

$$+ v'_{s,j}v'_{p,k}\langle A_{p\to s,i}|\mathbf{Z}_s = \mathbf{z}_s\rangle \Big] F'^{E}_{s}\,\mathrm{d}\mathbf{v}'_p\mathrm{d}\mathbf{v}'_s\mathrm{d}\delta'_p. \qquad (5.14)$$

## 5.2.5 Mixed Triple Correlation $\left\langle \phi'_p u'_{s,i} u'_{p,j} \right\rangle$

The mixed velocity-diameter triple correlation $\left\langle \phi'_p u'_{s,i} u'_{p,j} \right\rangle$ transport equation is obtained by setting $\mathscr{H}'_p = \phi'_p u'_{s,i} u'_{p,j}$ as follows

$$\alpha_p\rho_p\frac{\mathrm{D}_p}{\mathrm{D}t}\Big[\big\langle \phi'_p u'_{s,i} u'_{p,j}\big\rangle\Big] = -\frac{\partial}{\partial x_k}\Big[\alpha_p\rho_p\big\langle \phi'_p u'_{s,i} u'_{p,j} u'_{p,k}\big\rangle\Big]$$

$$+ \alpha_p\rho_p\big\langle A_{s,i}\phi'_p u'_{p,j} + A_{p,j}\phi'_p u'_{s,i}\big\rangle$$

$$- \alpha_p\rho_p\Bigg[\frac{\mathrm{D}_p\langle u_{s,i}\rangle}{\mathrm{D}t}\big\langle \phi'_p u'_{p,j}\big\rangle + \frac{\mathrm{D}_p\langle u_{p,j}\rangle}{\mathrm{D}t}\big\langle \phi'_p u'_{s,i}\big\rangle$$

$$+ \frac{\mathrm{D}_p\langle \phi_p\rangle}{\mathrm{D}t}\big\langle u'_{s,i} u'_{p,j}\big\rangle\Bigg]$$

$$- \alpha_p\rho_p\Bigg[\big\langle \phi'_p u'_{p,j} u'_{p,k}\big\rangle\frac{\partial\langle u_{s,i}\rangle}{\partial x_k} + \big\langle \phi'_p u'_{s,i} u'_{p,k}\big\rangle\frac{\partial\langle u_{p,j}\rangle}{\partial x_k}$$

$$+ \big\langle u'_{s,i} u'_{p,j} u'_{p,k}\big\rangle\frac{\partial\langle \phi_p\rangle}{\partial x_k}\Bigg] + S^{\langle \phi'_p u'_{s,i}u'_{p,j}\rangle}_{p\to s,ijk},$$

$$(5.15)$$

where

$$S^{\langle \phi'_p u'_{s,i}u'_{p,j}\rangle}_{p\to s,ijk} \equiv \int\limits_{-\infty}^{+\infty} \delta'_p v'_{p,j}\langle A_{p\to s,i}|\mathbf{Z}_s = \mathbf{z}_s\rangle\, F'^{E}_{s}\,\mathrm{d}\mathbf{v}'_p\mathrm{d}\mathbf{v}'_s\mathrm{d}\delta'_p. \qquad (5.16)$$

## 5.2.6 Mixed Triple Correlation $\left\langle \phi'_p \phi'_p u'_{p,i} \right\rangle$

The mixed velocity-diameter triple correlation $\left\langle \phi'_p \phi'_p u'_{p,i} \right\rangle$ transport equation is obtained by setting $\mathscr{H}'_p = \phi'_p \phi'_p u'_{p,i}$ as follows:

$$\alpha_p \rho_p \frac{D_p}{Dt} \left[ \left\langle \phi_p' \phi_p' u_{p,i}' \right\rangle \right] = -\frac{\partial}{\partial x_j} \left[ \alpha_p \rho_p \left\langle \phi_p' \phi_p' u_{p,i}' u_{p,j}' \right\rangle \right]$$

$$+ \alpha_p \rho_p \left\langle A_{p,i} \phi_p' \phi_p' \right\rangle$$

$$- \alpha_p \rho_p \left[ \frac{D_p \left\langle u_{p,i} \right\rangle}{Dt} \left\langle \phi_p' \phi_p' \right\rangle + 2 \frac{D_p \left\langle \phi_p \right\rangle}{Dt} \left\langle \phi_p' u_{p,i}' \right\rangle \right]$$

$$- \alpha_p \rho_p \left[ \left\langle \phi_p' \phi_p' u_{p,j}' \right\rangle \frac{\partial \left\langle u_{p,i} \right\rangle}{\partial x_j} + 2 \left\langle \phi_p' u_{p,i}' u_{p,j}' \right\rangle \frac{\partial \left\langle \phi_p \right\rangle}{\partial x_j} \right].$$

$$(5.17)$$

### 5.2.7 Particle Diameter Triple Correlation $\left\langle \phi_p' \phi_p' \phi_p' \right\rangle$

The particle diameter triple correlation $\left\langle \phi_p' \phi_p' \phi_p' \right\rangle$ transport equation is obtained by setting $\mathcal{H}_p' = \phi_p' \phi_p' \phi_p'$ as follows:

$$\alpha_p \rho_p \frac{D_p}{Dt} \left[ \left\langle \phi_p' \phi_p' \phi_p' \right\rangle \right] = -\frac{\partial}{\partial x_i} \left[ \alpha_p \rho_p \left\langle \phi_p' \phi_p' \phi_p' u_{p,i}' \right\rangle \right]$$

$$- 3 \alpha_p \rho_p \left\langle \phi_p' \phi_p' u_{p,i}' \right\rangle \frac{\partial \left\langle \phi_p \right\rangle}{\partial x_i} - 3 \alpha_p \rho_p \frac{D_p \left\langle \phi_p \right\rangle}{Dt} \left\langle \phi_p' \phi_p' \right\rangle.$$

$$(5.18)$$

The $\left\langle \phi_p' \phi_p' \phi_p' \right\rangle$ transport equation is particularly important for a polydispersed description of the flow. For example, particles with an initially Gaussian particle size distribution which are injected into a flow will undergo a distortion to the PDF of particle size due to particle drag. In order to capture the skewness of the local size PDF, the particle size triple correlation will be required.

## 5.3   Non-integer Closure Problem

Terms containing the drag force $A_{p,i}$ appear in Eqs. (4.60), (4.68), (4.70) and (4.73). These terms depending on the form of the drag law used, generate complicated non-integer moments. Considering Eq. (4.60) which is the simplest case, $\left\langle A_{p,i} \right\rangle$ can be expanded for a polydispersed particle population as follows

$$\left\langle A_{p,i} \right\rangle = \left\langle \frac{1}{\tau_p} (u_{s,i} - u_{p,i}) \right\rangle, \qquad (5.19)$$

where the particle response time is given by

$$\tau_p = \frac{\rho_p}{18\mu_f} \frac{\phi_p^2}{f_1},$$
(5.20)

and $f_1 = 1$ for Stokes flow. Substituting Eq. (5.20) into Eq. (5.19) and assuming Stokes flow results in

$$\langle A_{p,i} \rangle = \frac{18\mu_f}{\rho_p} \left\langle u_{s,i}\phi_p^{-2} - u_{p,i}\phi_p^{-2} \right\rangle.$$
(5.21)

If the flow were monodispersed the particle diameters would be independent of integration variables and Eq. (5.21) would simplify to

$$\langle A_{p,i} \rangle = \frac{18\mu_f}{\rho_p\phi_p^2} \langle u_{s,i} - u_{p,i} \rangle,$$
(5.22)

and no further closures were required. If the assumption of Stokes drag (for $Re_p > 1$) is to be further relaxed, a value for the parameter $f_1$ should be prescribed. A common relation in terms of particle Reynolds number for $f_1$ is given by the following equation [2]

$$f_1 = 1 + 0.15 Re_p^{0.687}.$$
(5.23)

Now substituting Eq. (5.23) into Eq. (5.19) results in

$$\langle A_{p,i} \rangle = \frac{18\mu_f}{\rho_p} \left[ \left\langle (u_{s,i} - u_{p,i})\phi_p^{-2} \right\rangle \right.$$

$$\left. + 0.15 \left( \frac{\rho_f}{\mu_f} \right)^{0.687} \left\langle \phi_p^{-1.313} \| u_{s,i} - u_{p,i} \|^{0.687} (u_{s,i} - u_{p,i}) \right\rangle \right],$$
(5.24)

which contains non-integer moments. Even more complicated closures are generated in Eqs. (4.68), (4.70) and (4.73) and closure of these moments in terms of available moments is not trivial. It is required for any proposed closure to be consistent with the presented stochastic framework and also to maintain the generality of the approach. If the form of the PDF is known any moment can directly be calculated by simple integrations over the state space. However, the only information available after solving the transport equations is a set of integer moments. The reconstruction of a PDF from a set of moments and also other possible approaches is the subject of the next chapter.

## 5.4 Summary

In this section, closure of different triple correlations encountered in the current EE field equations is considered by providing their transport equations and although no actual model is provided, it is shown that this approach is the most suitable route to writing closures for these terms. It is also discussed that the methods utilised to tackle the single phase triple correlation problem can be a good starting point. Then the non-integer moment problem is discussed and the PDF reconstruction methods are identified as a promising approach to tackle the non-integer and negative moment problem.

## References

1. Hanjalic K, Launder BE (1972) A reynolds stress model of turbulence and its application to thin shear flows. J Fluid Mech 52:609–638
2. Zaichik LI, Alipchenkov V (2001) A statistical model for transport and deposition of high-inertia colliding particles in turbulent flow. Int J Heat Fluid Flow 22:365–371

# Chapter 6
# PDF Reconstruction Methods

In this chapter, the closure problem for non-integer moments as discussed in Sect. 5.3 will be considered. In this chapter, only one-dimensional moments of form $\langle x^p \rangle$, $p \in \mathbb{R}$, are considered. It is worth mentioning that although the closure problem discussed in Sect. 5.3 is a multi-variate problem in nature, this preliminarily one-dimensional treatment can conceivably provide insight into the general multivariate case. In addition, the univariate case is encountered in multiphase flow systems if heat and mass transfer, chemical reactions, agglomeration, or break-up are added to the equations, and hence, the problem is also of practical interest. Two different categories of methods are considered: the first method is based on the reconstruction of the underlying probability density function (PDF) using Laguerre polynomials and the other is based on the direct calculation of non-integer moments using the fractional derivatives of moment-generating function (MGF). By applying the results of fractional calculus, an explicit equation is derived to express non-integer moments as a function of any arbitrary number of integer moments. The proposed methods are tested on several highly non-Gaussian analytical PDFs in addition to experimental agglomeration data and direct numerical simulation of fluid–particle turbulent multiphase flows.

## 6.1 Introduction

The governing equations for many physical phenomena are precisely known; however, numerical simulation by direct discretization of the governing equations is not generally feasible when several complicated phenomena such as turbulence, chemical reaction, agglomeration and break-up are involved. The starting point to tackle such complicated phenomena is usually to write the averaged equations. The process starts by defining an arbitrary PDF for the process and deriving an equation for the evolution of the PDF as discussed in Chap. 4. Rigopoulos [1] also discusses similar methods from an engineering point of view using the population balance equation (PBE) for reactive flows.

J. S. Shrimpton et al., *Statistical Treatment of Turbulent Polydisperse Particle Systems*, Green Energy and Technology, DOI: 10.1007/978-1-4471-6344-2_6, © Springer-Verlag London 2014

Non-integer moments can be introduced in the equations if the size distribution of particles is retained in Eulerian field equations (moment evolution) as discussed in Chap. 5. Also, if phenomena such as coagulation, nucleation, coalescence and break-up are included in the PBE [2–6] or heat and mass transfer considered [7, 8], they are present. There are some ad hoc solutions to the problem, such as interpolation between the moments. Beck and Watkins [7, 8] used geometric interpolation, and Frenklach and Harris [2] used Laguerre interpolation between the logarithms of the integer moments. Note that interpolation between the moments is not possible for negative fractional moments since usually only equations for positive moments are available. Fractional negative moments can also appear in the moment evolution equations as discussed in Sect. 5.3. PDF reconstruction methods are also attempted [3–5], which will be discussed in more detail below.

In this chapter, the problem is examined from a mathematical point of view and the moments $\mu_p = \langle x^p \rangle$ of a general univariate PDF are considered. The moments are given by

$$\mu_p = \int_\Omega x^p \mathscr{P}(x) \mathrm{d}x, \quad p \in \mathbb{R}, \tag{6.1}$$

where $\Omega$ is the domain on which $x$ is defined. It is obvious that by having the PDF, any moment can readily be calculated; however, in many physical phenomena, it is easy to determine the moments, but it is extremely difficult to determine the distributions themselves [9]. In addition, during a numerical simulation, the only information available are the solved variables, and in this case, the solved variables are the first few integer moments not the evolution equation of the PDF itself. Thus, a method is required to enable us to estimate the real-order moments $\mu_p, p \in \mathbb{R}$, using limited number of integer moments $\mu_i$, $i = 1, 2, \ldots, n$. To accomplish this, two general methods will be contemplated: a PDF reconstruction method based on Laguerre polynomials and a direct fractional method of moments (DFMM) based on derivatives of the MGF.

The first obvious approach is to try to reconstruct the PDF using integer moments $\mu_i$, which leads to the well-known finite-moment problem [9]. The finite-moment problem can be regarded as a finite-dimensional version of the Hausdorff moment problem [10–13], which is in general ill-posed lacking one or more conditions of a well-posed problem (i.e. existence, uniqueness and stability) [14]. The reconstruction methods can be classified under linear and non-linear methods as discussed by Volpe and Baganoff [15]. They classified the maximum entropy method (MEM) as a non-linear method and reconstruction methods based on expansion around some parent distribution as linear methods.

The MEM was first utilized by Koopman [16] based on the idea of information theory of Shannon [17]. The MEM (Koopman) method is attractive for several reasons [18]: (i) sound conceptual foundations; (ii) interdependence between even- and odd-order moments; (iii) non-negative probabilities; and (iv) produces the most unbiased distribution possible, i.e. if a distribution with less entropy (uncertainty) were used, that would imply the existence of additional knowledge [19]. The last property of MEM (i.e. most unbiased distribution) might be interesting in some applications, but for the current problem where only the first few moments (usually 2) are available, the method effectively only generates Gaussian PDFs. However, since the method has been suggested for poly-dispersed particulate flow, it will be succinctly discussed in this chapter.

PDF reconstruction based on some priori simple shape is the other option. However, this method has a global realizability issue, in that an assumed PDF, valid at one location in space and time, may evolve into a form that violates the assumption of the PDF at another position [8]. A more advanced method in this category is expansion using orthogonal polynomials. Two basic methods in this category that are extensively used in the literature are Gram–Charlier and Edgeworth series expansion [20]. In both methods, the PDF $\mathscr{P}$ is evaluated using a truncated expansion in terms of Hermite polynomials $H_n(x)$:

$$\mathscr{P}(x) = \sum_{n=1}^{N} C_n H_n(x) \mathscr{N}(\mu, \sigma), \qquad (6.2)$$

where $\mathscr{N}(\mu, \sigma)$ is a Gaussian distribution with parameters $\mu$ and $\sigma$ and $C_n$ are the coefficients containing the higher-order moments (Gram–Charlier) or higher-order cumulants (Edgeworth). $H_n$ is the Hermite polynomial which can explicitly be written [21] by

$$H_n(x) = n! \sum_{k=0}^{n/2} \frac{(-1)^k x^{n-2k}}{k!(n-2k)!2^k}. \qquad (6.3)$$

Majumdar et al. [22] among others [23, 24] discussed that both methods can be divergent due to the fact that the series expansion is sensitive to the behavior of $\mathscr{P}(x)$ at infinity. For the series to be convergent, $\mathscr{P}(x)$ should fall faster to zero than $e^{-x^2/4}$. Generalized Laguerre polynomial expansion is the more promising approach in this category, which does not have any oscillatory or divergence problems of Edgeworth or Gram–Charlier expansions [22]. Although Laguerre polynomial expansion has been used in the literature to reconstruct different types of PDFs, its accuracy in estimating non-integer moments encountered in multiphase flow systems has not been tested and will be discussed in this chapter. There are other methods that cannot be considered general such as spline reconstruction method proposed by John et al. [9] which fits cubic splines to the available moments; however, for the method to be effective, usually a large number of moments are required.

The methods discussed above are all indirect methods, i.e. first, a PDF is reconstructed using one of the discussed methods, and then, the PDF should numerically be integrated [Eq. (6.1)] to calculate the required moments. In this chapter, a direct method using fractional derivatives of the MGF is formulated based on the work of Haeri and Shrimpton [25]. The property of the moment-generating function is that its $k$th derivative calculated at $s = 0$ is equal to the $k$th moment of the distribution $\mathscr{P}(x)$ [26]:

$$\mu_k = \left. \frac{\mathrm{d}^k G}{\mathrm{d} s^k} \right|_{s=0}. \tag{6.4}$$

Now, if this equation could be extended to non-positive non-integer values of $k$, then it would provide a way of expressing non-integer moments as a function of MGF. This extension is formalized in the field of fractional calculus and will be discussed in Sect. 6.4. Then, several examples will be provided using moments of analytical PDFs.

## 6.2 Maximum Entropy Method

Maximum entropy method was first utilized by Koopman [16] based on the idea of information theory of Shannon [17]. The MEM (Koopman) method is attractive for several reasons: (i) sound conceptual foundations [18]; (ii) interdependence between even- and odd-order moments [27]; (iii) non-negative probabilities; and (iv) produces the most unbiased distribution possible, i.e. if a distribution with less entropy (uncertainty) were used, that would imply the existence of additional knowledge [19, 28]. MEM is also used in the spray atomization community for predicting droplet size–velocity distributions [29–33] where input constraints are generally formulated in terms of atomizer characteristics and conservation of energy, mass and momentum for the droplet phase. However, the models developed are generally predictive methods for steady conditions and are not particularly relevant to the application proposed here, namely closure of acceleration terms in the transport model. An application of the method to similar problem can be found in [19, 34], but despite proposing the MEM to close the equations, Archambault chose not to implement the closure in favour of an analytical Gaussian. This decision is actually a clever one, since having the first two moments, MEM will always result in a Gaussian distribution (see Sect. 6.2.1). Therefore, for the EE models where commonly only the first two moments are available, there is no point wasting computational time on MEM.

The entropy is defined for a group of $N$ mutually exclusive events, each with probability $p_i$. Shannon's measure of the entropy or uncertainty of the distribution is

$$H = -C \sum_{i=1}^{N} p_i \ln p_i, \tag{6.5}$$

where $C$ is an arbitrary positive constant. Given a continuous distribution in one dimension $f(x)$, the entropy of the distribution can be defined as the functional

$$H[f(x)] = - \int_{-\infty}^{+\infty} f(x) \ln[f(x)] \mathrm{d}x, \tag{6.6}$$

Subject to

$$J_0[f(x)] = \int\limits_{-\infty}^{+\infty} f(x)\mathrm{d}x = 1, \tag{6.7}$$

$$J_i[f(x)] = \int\limits_{-\infty}^{+\infty} g_i(x)f(x)\mathrm{d}x = \langle g_i(x) \rangle, \tag{6.8}$$

where $J_0$ is the normalization constraint and $J_1$ is what we require (i.e. for $f$ to be a evolution PDF of $g$). Formulating the equations as an optimization problem, using Lagrange multipliers, we have [35, 36]

$$\left\{ \frac{\partial}{\partial f} - \frac{\mathrm{d}}{\mathrm{d}x}\frac{\partial}{\partial f'} \right\} \left[ -f \ln[f] + \lambda_0 f + \sum_{n=1}^{N} \lambda_n g_n(x) f \right] = 0, \tag{6.9}$$

which can be solved to (also swapping the sign of Lagrange parameters for convenience) get

$$f(x) = \exp\left[ -\left( \lambda_0 + \sum_{n=1}^{N} \lambda_n g_n(x) \right) \right]. \tag{6.10}$$

Equation (6.10) demonstrates that the PDF is completely specified by the parameters $\lambda_0$ and $\lambda_n$, which in turn are associated with each of the constraints $\langle g_n(x) \rangle$. Ideally, it would be possible to find analytical solutions of the form $\lambda_n = \lambda_n[\langle g_m(x) \rangle]$ by substituting Eq. (6.10) into Eqs. (6.8) and (6.7); however, due to non-linear nature of the equations, a numerical procedure is inevitable (see [37, 38] for an implementation). Extension to multidimensional PDFs is trivial, and entropy $H[f(\mathbf{x})]$ and constraints $J_i[f(\mathbf{x})]$ simply become

$$H[f(\mathbf{x})] = -\int\limits_{-\infty}^{+\infty} f(\mathbf{x}) \ln[f(\mathbf{x})]\mathrm{d}\mathbf{x}, \tag{6.11}$$

$$J_0[f(\mathbf{x})] = \int\limits_{-\infty}^{+\infty} f(\mathbf{x})\mathrm{d}\mathbf{x} = 1, \tag{6.12}$$

$$J_i[f(\mathbf{x})] = \int\limits_{-\infty}^{+\infty} g_i(\mathbf{x})f(\mathbf{x})\mathrm{d}\mathbf{x} = \langle g_i(\mathbf{x}) \rangle, \tag{6.13}$$

where the integrals are taken over the entire multidimensional phase space. Following a similar reasoning, the multidimensional PDF is given as

$$f(\mathbf{x}) = \exp\left[-\left(\lambda_0 + \sum_{i=1}^{N} \lambda_n g_n(\mathbf{x})\right)\right]. \tag{6.14}$$

### 6.2.1  Simple Analytical Solutions Using MEM

In order to demonstrate the validity of the MEM, it is useful to consider some simple one-dimensional examples that can be solved analytically. First, consider the case where only the normalization constraint is imposed. In this situation,

$$f(x) = e^{-\lambda_0}, \tag{6.15}$$

which corresponds to a uniform distribution, which in general form is

$$f(x) = \begin{cases} e^{-\lambda_0} & \text{for } |x| \le a, \\ 0 & \text{for } |x| > a, \end{cases} \tag{6.16}$$

and the normalization constraint gives $\lambda_0 = \ln(2a)$ with $H[f(x)] = \ln(2a)$. This results in

$$f(x) = \begin{cases} \frac{1}{2a} & \text{for } |x| \le a, \\ 0 & \text{for } |x| > a. \end{cases} \tag{6.17}$$

If the mean is added to the constraints, the distribution function becomes

$$f(x) = e^{-(\lambda_0 + \lambda_1 x)}. \tag{6.18}$$

Further evaluation of the constraint integrals leads to $\lambda_0 = \ln(\langle x \rangle)$, $\lambda_1 = \langle x \rangle^{-1}$, and the entropy $H[f(x)] = 1 - \ln(\langle x \rangle)$. These relations give the exponential PDF in traditional form

$$f(x) = \frac{1}{\langle x \rangle} e^{x/\langle x \rangle} \tag{6.19}$$

Finally, the Gaussian PDF is obtained if the zeroth-, first- and second-order moments are constrained to give

$$f(x) = e^{-(\lambda_0 + \lambda_1 x + \lambda_2 x^2)}, \tag{6.20}$$

and the Lagrange parameters are evaluated as

$$\lambda_0 = \frac{\langle x \rangle^2}{2\sigma^2} + \frac{1}{2} \ln(2\pi\sigma^2), \tag{6.21}$$

$$\lambda_1 = -\frac{\langle x \rangle}{\sigma^2},$$

$$\lambda_2 = -\frac{1}{2\sigma^2}.$$

Finally, the familiar form of the Gaussian is realized

$$f(x) = \frac{1}{\sqrt{2\pi\sigma^2}} \exp\left[-\frac{(x - \langle x \rangle)^2}{2\sigma^2}\right]. \tag{6.22}$$

The maximum entropy formulation for Gaussian distributions is revealing as it says that maximal ignorance with regard to moments higher than two implies the Gaussian distribution.

## 6.3 Generalized Laguerre Polynomial Expansion

The Laguerre expansion of a PDF $\mathscr{P}$ can be written by Majumdar et al. [22] and Mustapha and Dimitrakopoulos [39]

$$\mathscr{P}(x) = \sum_{n=0}^{\infty} r_n L_n^a(bx) \mathscr{G}(x; a + 1, b), \tag{6.23}$$

where $L_n^a$ is the generalized Laguerre polynomial with parameter $a$. $\mathscr{G}(x; a, b)$ is the gamma distribution with parameters $a$ and $b$ given by

$$\mathscr{G}(x; a, b) = \frac{b^a}{\Gamma(a)} x^{a-1} \exp(-bx), \quad x \geq 0, \ a, b > 0. \tag{6.24}$$

In Eq. (6.24), $\Gamma(a)$ is the gamma function given by

$$\Gamma(a) = \int_0^{\infty} t^{a-1} \exp(-t) dt. \tag{6.25}$$

If the argument of the gamma function is an integer, then $\Gamma(a) = (a - 1)!$. The function can be approximated by directly calculating the integral which is not a stable method, and a better approach would be the approximation method proposed by Cody [40]. Generalized Laguerre polynomials are orthogonal with respect to the measure $\mathscr{G}$; therefore, using the Rodriguez formula [41], $L_n$ can be defined by

$$L_n^a(x)\mathscr{G}(x; a + 1, 1) = \frac{1}{n!} \left(\frac{d}{dx}\right)^n [x^n \mathscr{G}(x; a + 1, 1)]. \tag{6.26}$$

Using the Leibniz's theorem for differentiation of product, it can explicitly be written by Lebedev [41]

$$L_n^a(x) = \sum_{m=0}^{n} \binom{n + a}{n - m} \frac{(-x)^m}{m!}, \tag{6.27}$$

where the binomial coefficients are generalized using the gamma function.

The coefficients $r_n$ can be found using the orthogonality of the Laguerre polynomials. A sequence of polynomials $Q_1(x), \ldots, Q_n(x)$ are said to be orthogonal on
the interval $[x_1, x_2]$ with respect to a weight function $W(x)$ if

$$\int_{x_1}^{x_2} Q_n Q_m W(x) \mathrm{d}x = 0 \quad \forall m \neq n. \tag{6.28}$$

For Laguerre polynomials, it can easily be shown that

$$\int_0^\infty L_n^a(bx) L_m^a(bx) \mathscr{G}(x; a+1, b) \mathrm{d}x = \frac{\Gamma(n+a+1)}{n! \Gamma(a+1)} \delta_{nm}, \tag{6.29}$$

where $\delta_{nm} = 1$ if $m = n$ and is 0 otherwise.

The coefficients $r_n$ of the Laguerre expansion can be found by multiplying right-
and left-hand sides of Eq. (6.23) by $L_m^a(x)$ and using Eq. (6.29), which results in

$$r_n = \frac{n! \Gamma(a+1)}{\Gamma(n+a+1)} \int_0^\infty \mathscr{P}(x) L_n^a(bx) \mathrm{d}x. \tag{6.30}$$

Introducing Eq. (6.27) into Eq. (6.30), we have

$$r_n = \frac{n! \Gamma(a+1)}{\Gamma(n+a+1)} \int_0^\infty \mathscr{P}(x) \sum_{i=0}^n \binom{n+a}{n-i} \frac{(-bx)^i}{i!} \mathrm{d}x$$

$$= \frac{n! \Gamma(a+1)}{\Gamma(n+a+1)} \sum_{i=0}^n \binom{n+a}{n-i} \frac{(-b)^i}{i!} \int_0^\infty \mathscr{P}(x) x^i \mathrm{d}x$$

$$= \frac{n! \Gamma(a+1)}{\Gamma(n+a+1)} \sum_{i=0}^n \binom{n+a}{n-i} \frac{(-b)^i m_i}{i!}$$

$$= n! \Gamma(a+1) \sum_{i=0}^n \frac{(-b)^i m_i}{i!(n-i)! \Gamma(a+i+1)}$$

In this equation, $\mu_i$ are the available moments of order $i$. The only remaining issue is
the calculation of the parameters $a$ and $b$ which are defined such that the expansion
(6.23) has the same mean and variance as the original distribution $\mathscr{P}(x)$. Two parameters of the Laguerre expansion can be found by requiring that the first and second
moments of the fitted distribution be equal to the first two moments of the original
distribution, i.e.

$$\int_0^\infty \mathscr{P}(x)L^a(\beta x)\mathrm{d}x = 0 \quad \text{for } n = 1, 2. \tag{6.31}$$

Using Eq. (6.27) in Eq. (6.31) and expanding for $n = 1$ and $n = 2$, we have

$$\phi(x) = \begin{cases} a - 1 + bm_1 = 0, & n = 1 \\ \frac{(a+1)(a+2)}{2} - (a+2)bm_1 + b^2\frac{m_2}{2} = 0, & n = 2, \end{cases} \tag{6.32}$$

which can be solved for $a$ and $b$ to get [39]

$$a = \frac{2m_1^2 - m_2}{m_2 - m_1^2}, \quad \text{and} \quad b = \frac{m_1}{m_2 - m_1^2}. \tag{6.33}$$

## 6.4 Direct Fractional Method of Moments

In this section, an approach to estimate fractional moments as a series of the integer moments is formulated. The MGF of a positive-valued distribution can be defined by [26]:

$$G(s) = \int_0^\infty \mathscr{P}(u)e^{su}\mathrm{d}u. \tag{6.34}$$

The main property of the MGF is that the moments of the original PDF $\mathscr{P}(x)$ can be derived from this function by Eq. (6.4). Extension of this equation to non-integer values will follow next.

### 6.4.1 Fractional Derivatives and Integrals

The derivative of arbitrary real order $p$ can be considered as an interpolation of the operators of a sequence of n-fold integration and $n$th-order derivative. Fractional derivatives are presented by $_aD_t^p(t)$ where $a$ and $t$ are the limits related to the operation of the fractional differentiation and are commonly called terminals of the fractional differentiation. These are essential to avoid ambiguities in application of fractional derivative to real numbers [42–44]. There are two equivalent approaches to the definition of fractional differentiation, namely the Grunwald–Letnikov (GL) approach and the Riemann–Liouville (RL) approach. It is customary [43] to use the RL formulation for problem set-up and use GL approach to obtain a numerical solution. Following the same approach, RL definition is used to establish the relation between the evaluation of the moments and fractional calculus, and GL approach is used to provide an explicit equation for fractional moments as a function of integer moments.

### 6.4.2 RL Fractional Derivative

Consider the integral

$$f^{-1}(t) = \int_a^t f(\tau)d\tau, \tag{6.35}$$

and define

$$f^{-2}(t) = \int_a^t d\tau_1 \int_a^{\tau_1} f(\tau)d\tau. \tag{6.36}$$

Now, by setting $h(\tau_1) = \int_a^{\tau_1} f(\tau)d\tau$ and $g'(\tau_1) = 1$, we have

$$\int_a^t d\tau_1 \int_a^{\tau_1} f(\tau)d\tau = \int_a^t h(\tau_1)g'(\tau_1)d\tau_1 = h(\tau_1)g(\tau_1)|_a^t - \int_a^t \tau_1 f(\tau_1)d\tau_1$$

$$= t\int_a^t f(\tau)d\tau - \int_a^t \tau f(\tau)d\tau = \int_a^t (t - \tau)f(\tau)d\tau. \tag{6.37}$$

It is now easy to use Eq. (6.37) $n$ times to show by induction that Eq. (6.38) holds for any integer $n$:

$$f^{-n}(t) = \frac{1}{(n-1)!}\int_a^t (t - \tau)^{n-1} f(\tau)d\tau, \quad n \geq 1. \tag{6.38}$$

Equation (6.38) can be extended to non-integer values of $n$ using the gamma function:

$$_aD_t^{-p}f(t) = \frac{1}{\Gamma(p)}\int_a^t (t - \tau)^{p-1} f(\tau)d\tau, \quad p > 0. \tag{6.39}$$

The fractional derivative can then be defined by Schiavone and Lamb [45] and Cottone and Di Paola [46]

$$_aD_t^p f(t) = \frac{1}{\Gamma(k-p)} \left(\frac{d}{dt}\right)^k \int_a^t (t-\tau)^{k-p-1} f(\tau)d\tau, \quad k-1 \le p \le k. \quad (6.40)$$

Note that this definition is not arbitrary and it is defined to be equivalent to the GL definition which is defined as an extension to backward differences. First note that using Eq. (6.40) with $v = k - p$, we have

$$_{-\infty}D_t^p f(t) = \frac{1}{\Gamma(v)} \left(\frac{d}{dt}\right)^k \int_{-\infty}^t (t-\tau)^{v-1} f(\tau)d\tau. \quad (6.41)$$

Using $\eta = t - \tau$,

$$_{-\infty}D_t^p f(t) = \frac{1}{\Gamma(v)} \left(\frac{d}{dt}\right)^k \int_0^\infty \eta^{v-1} f(t-\eta)d\eta$$

$$= \frac{1}{\Gamma(v)} \int_0^\infty \eta^{v-1} \left(\frac{d}{dt}\right)^k f(t-\eta)d\eta$$

$$= \frac{1}{\Gamma(v)} \int_{-\infty}^t (t-\tau)^{v-1} \left(\frac{d}{dt}\right)^k f(\tau)d\tau. \quad (6.42)$$

The last line of Eq. (6.42) is the Caputo [47] definition of the fractional derivative with $a \to -\infty$, i.e. $_{-\infty}^C D_t^p f(t)$. The RL and Caputa definitions are not generally equivalent; however, Eq. (6.42) shows that both definitions are equivalent for the limit $a \to -\infty$. Now, we can use [48]

$$_{-\infty}^C D_t^p e^{ct} = c^p e^{ct} \quad \forall t, \; -\infty < t < \infty, \; \forall c > 0, \quad (6.43)$$

and consequently, $_{\infty}D_t^p e^{ct} = c^p e^{ct}$. Assuming $G(s)$ is analytic in $[-\infty, 0]$ and defining $v = k - p$, one ca use Eq. (6.40) and change the order of integration to get

$$_{-\infty}D_s^p G(s)|_{s=0} = \Gamma(v)^{-1} \frac{d^k}{ds^k} \int_{-\infty}^s (s-\tau)^{v-1} \left[\int_0^\infty e^{\tau u} \mathscr{P}(u)du\right] d\tau \Bigg|_{s=0}$$

$$= \int_0^\infty \left[\Gamma(v)^{-1} \frac{d^k}{ds^k} \int_{-\infty}^s (s-\tau)^{v-1} e^{\tau u}d\tau\right] \Bigg|_{s=0} \mathscr{P}(u)du$$

$$= \int_0^\infty {}_{-\infty}D_s^p (e^{su}) \Bigg|_{s=0} \mathscr{P}(u)du = \int_0^\infty u^p e^{su} \Bigg|_{s=0} \mathscr{P}(u)du$$

$$= \int_0^\infty u^p \mathscr{P}(u) \mathrm{d}u = \mu_p. \tag{6.44}$$

In Eq. (6.44), $\mu_p$ is the moment of the density $\mathscr{P}(x)$ of order $p$. Although the RL definition of fractional derivatives and integrals is more convenient to mathematically link the fractional moments to fractional derivatives, it does not provide a numerical method for calculating the moments. However, by reverting to the GL definition, an expression can be written for the fractional moments.

### 6.4.3  GL Fractional Derivatives

GL definition of fractional derivatives is more intuitive and starts by observing the series of backward differences and writing a general series for derivatives of order $n$:

$$f^{(n)}(t) = \frac{\mathrm{d}^n f}{\mathrm{d}t^n} = \lim_{h \to 0} \frac{1}{h^n} \sum_{r=0}^n (-1)^r \binom{n}{r} f(t - rh). \tag{6.45}$$

Then, by considering the following generalization,

$$f_h^{(p)}(t) = \frac{1}{h^p} \sum_{r=0}^n (-1)^r \binom{p}{r} f(t - rh), \tag{6.46}$$

for arbitrary natural numbers $p$ and $n$, such that $p \le n$, we have

$$\lim_{h \to 0} f_h^p(t) = f^{(p)}(t) = \frac{\mathrm{d}^p f}{\mathrm{d}t^p}, \tag{6.47}$$

because all the coefficients in the numerator after $\binom{p}{p}$ are identically zero. Equation (6.46) can be inverted by expanding the series for the first few terms:

$$f^{(2)}(t) = h^{-2} f(t) - 2h^{-2} f(t - h) + h^{-2} f(t - 2h) \tag{6.48}$$

$$f^{(1)}(t) = h^{-1} f(t) - h^{-1} f(t - h) \tag{6.49}$$

$$f^{(0)}(t) = f(t). \tag{6.50}$$

Then, by introducing Eq. (6.50) into Eq. (6.49) to eliminate $h^{-1} f(t)$, we get

$$f(t - h) = f(t) - h f^{(1)}(t), \tag{6.51}$$

similarly by introducing Eqs. (6.50) and (6.49) into Eq. (6.48) to eliminate first two terms on the RHS, we have

$$f(t - 2h) = f(t) - 2hf^{(1)}(t) + h^2 f^{(2)}(t). \tag{6.52}$$

The following equation then follows easily by induction

$$f(t - ih) = \sum_{r=0}^{i} (-1)^r \binom{i}{r} h^r f^{(r)}(t). \tag{6.53}$$

Equation (6.53) is the key for writing the non-integer moments as functions of integer moments, and it will become clear in the course of this section. To extend the definition to negative values, a careful definition of the upper bound of the summation is required; otherwise, the limit in Eq. (6.46) would be strictly zero for any $n$. Therefore, by taking $h = \frac{t-a}{n}$, $a$ being a real number, we can define

$$\lim_{\substack{h \to 0 \\ nh=t-a}} f_h^{(-p)}(t) = {}_a D_t^{-p} f(t). \tag{6.54}$$

Note that Eq. (6.54) is actually the definition of an integral; for example, writing the series for $p = 1$, the limit is simply the definition of $\int_a^t f(\tau) d\tau$. By the method of induction, it is possible to show [43, 44]

$$_a D_t^{-p} f(t) = \lim_{\substack{h \to 0 \\ nh=t-a}} h^p \sum_{r=0}^{n} (-1)^r \binom{p}{r} f(t - rh) = \frac{1}{(p-1)!} \int_a^t (t-\tau)^{p-1} f(\tau) d\tau, \tag{6.55}$$

and consequently, [43]

$$_a D_t^{-p} f(t) = \int_a^t d\tau_1 \int_a^{\tau_1} d\tau_2 \dots \underbrace{\int_a^{\tau_{p-1}} f(\tau_p) d\tau_p}_{p \text{ times}}. \tag{6.56}$$

Therefore,

$$_a D_t^{p} f(t) = \lim_{\substack{h \to 0 \\ nh=t-a}} h^{-p} \sum_{r=0}^{n} (-1)^r \binom{p}{r} f(t - rh), \tag{6.57}$$

is indeed a general expression for $p$-fold integration and derivatives of order $n$. Equation (6.57) can be extended to non-integer $p$ values. By direct calculation of the limits, it can be shown that [43, 46]

$$_a D_t^{-p} f(t) = \frac{1}{\Gamma(p)} \int_a^t (t-\tau)^{p-1} f(\tau) d\tau, \tag{6.58}$$

and

$$_a D_t^p f(t) = \sum_{r=0}^m \frac{f^{(r)}(a)(t-a)^{-p+r}}{\Gamma(-p+r+1)}$$

$$+ \frac{1}{\Gamma(-p+m+1)} \int_a^t (t-\tau)^{m-p} f^{(m+1)}(\tau) d\tau, \quad \forall m > p-1.$$

(6.59)

It is important to note the equivalence between Eqs. (6.39) and (6.58). The equivalence between Eqs. (6.59) and (6.40) is harder to note, but direct differentiation and integration by parts show that both definitions are indeed equivalent [44]. Having the framework established, an equation will be derived to explicitly write the non-integer moments as a function of integer moments in the next section.

### 6.4.4 Estimating the Non-integer Moments

A first-order approximation to $p$-order derivative using the GL definition can be written as

$$_a D_t^p f(t) = \lim_{h \to 0} h^{-p}{}_a \Delta_t^p f(t) \approx h^{-p} \sum_{r=0}^{\lfloor \frac{t-a}{h} \rfloor} (-1)^r \binom{p}{r} f(t-rh).$$

(6.60)

The number of addends in Eq. (6.60) becomes very large for $t \gg a$. This series can be truncated using the short-memory principle, taking into account the behavior of $f(t)$ only in the recent past. This means that the truncated series (6.60) is in particular a very good approximation for calculating fractions that are approximately equal to the number of integer moments retained in the expansion and the error can be quantified as suggested by Deng [49]. In Eq. (6.60), $f(t-rh)$ can be considered as the MGF, but since the function is not explicitly available, the equation is not usable in this form. Equation (6.53) provides an equation for this term, and now, inserting Eq. (6.53) into Eq. (6.60) and noting the definitions (6.44) and (6.4), we get

$$\mu_p \approx \sum_{r=0}^N (-1)^r \binom{p}{r} \sum_{j=0}^r (-1)^j \binom{r}{j} h^{j-p} \mu_j,$$

(6.61)

where $N$ is the number of moments retained in the expansion. A similar equation is recently derived by Gzyl and Tagliani [50] using Taylor expansions and also by Alexiadisa et al. [51] using Weyl fractional derivatives. However, they applied the series to larger number of integer moments ($O(100)$ and $O(10)$, respectively), which is not applicable to the problems considered in this study. The only remaining issues are to provide an equation to calculate the coefficient $w_r^p = (-1)^r \binom{p}{r}$ and propose a step size $h$. The coefficients, $w_r^p$, are generalizations of binomial coefficients to non-integer values and can be calculated using the definition of the gamma function. However, another possible approach is to use the recursive relations [43]

$$w_0^p = 1, \quad w_r^p = \left(1 - \frac{p+1}{r}\right) w_{r-1}^p, \quad r = 1, 2, \ldots, \tag{6.62}$$

which eliminates the need for the evaluation of gamma functions and provides better computational efficiency and stability. In addition, it provides a unique opportunity for higher-order estimations of the fractional derivative, which is discussed next. Considering the coefficients of power series for the function $(1 - z)^p$:

$$(1 - z)^p = \sum_{r=0}^{\infty} w_r^p z^r. \tag{6.63}$$

Substituting $z = e^{-i\theta}$ coefficients can be expressed in terms of Fourier transforms:

$$w_r^p = \frac{1}{2\pi i} \int_0^{2\pi} (1 - e^{-i\theta})^p e^{ik\theta} d\theta. \tag{6.64}$$

Since always only a finite number of moments are used, any FFT library can be used to calculate the coefficients using Eq. (6.64). Equations (6.62) and (6.64) are only first-order approximations; however, higher-order approximations can be constructed using higher-order polynomials. Lubich [52] suggests polynomials up to sixth-order which can be used in conjunction with Eq. (6.64) and an FFT library to find coefficients for the higher-order approximation to the fractional derivatives. However, only first-order approximation (Eq. 6.62) is used in this study due to its simplicity and efficiency.

The step size $h$ should ideally be very small. Prescribing a value for $H = h^{-1}$ is equivalent to the width of the PDF considered in the calculations. Gzyl and Tagliani [50] showed that the series is always convergent for $1/2 < h \le 1$ or $1 \le H < 2$; however, as will be discussed in the next section, this value causes a severe underestimation when only a few moments are used. It seems to be reasonable to relate $H$ to the mean and the variance of the available data:

$$H = \mu + \lambda\sigma, \tag{6.65}$$

where $\mu$ is the mean of the data, $\sigma$ is the standard deviation, and $\lambda$ is an adjustable parameter. Relating the step size to the statistics of the data was first explained in [43] and later used by Alexiadisa et al. [51] for the simulation of agglomeration problem using the method of moments (MOM) to solve the PBE; however, different forms with different parameters are possible. It was shown later in Haeri and Shrimpton [25] that if the fractional moment of interest is not much larger than the number of moments retained in the series, setting $\lambda = \frac{1}{p+1}$ provides very accurate estimates.

## 6.5 Examples

### 6.5.1 Log-Normal Distribution

To test the results, an analytic log-normal distribution given by

$$\mathscr{P}(x) = \frac{1}{\sqrt{2\pi}\sigma x} e^{-\frac{(\log x - \bar{\mu})^2}{2\bar{\sigma}^2}}, \tag{6.66}$$

is considered. In Eq. (6.66), $\bar{\sigma}$ and $\bar{\mu}$ are scale and location parameters, respectively. Higher-order moments are analytically given by

$$\mu_j = e^{(\frac{1}{2}j^2\sigma^2 + j\bar{\mu})}. \tag{6.67}$$

Figure 6.1 shows the simulation of the log-normal distribution using the Laguerre polynomials where the PDF is reconstructed for 200 equidistant points in range [0, 12]. Although our goal is to simulate the non-integer moments, the accuracy of the calculated moments is directly proportional to the accuracy of the fit. Generally, a Laguerre polynomial is able to produce a very good fit to this distribution with a limited number of moments because of the similarity between the shape of gamma and log-normal distributions.

Figure 6.2 shows the fractional moments $\mu_p$, $p = \{1.2, 2.3, 3.4, 4.5\}$ calculated using the fractional moments approach and the Laguerre polynomial approach. Intentionally, only the first two moments (normalized on [0, 12]) are used to test the ability of the two proposed methods in estimating the non-integer moments using only limited data. In this section, all errors are calculated by

$$\text{Err} = \frac{|\mu_a - \mu_e|}{\mu_a}, \tag{6.68}$$

where subscripts $a$ and $e$ stand for the analytical and estimated values, respectively, and also %Err = $100 \times$ Err. In Fig. 6.2, error bars, for $\pm 100 \times$ Err, are also provided with maximum error of 27.5 % in calculating the $\mu_{4.5}$ using the fractional moments method with all other errors being less than 5 %. The reason for this large error will be discussed in the next section, and a remedy will be suggested.

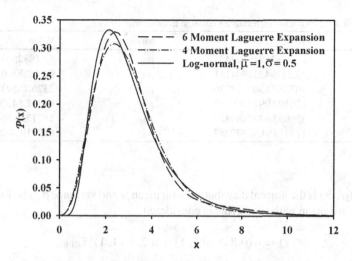

**Fig. 6.1** Log-normal distribution—scale and position parameters are 0.5 and 1, respectively. This distribution can accurately be reproduced using small number of moments

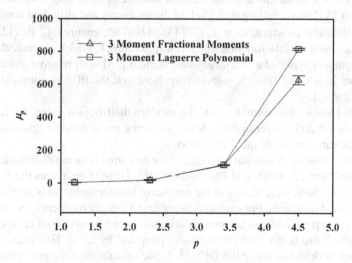

**Fig. 6.2** Non-integer moment estimation—LPM and DFMM methods are used, and error bars are presented for each calculation

## 6.5.2 Mixture of Normal Distributions

A mixture of normal distributions can be created using

$$\mathscr{P}(x) = \sum_{i=1}^{N} w_i \mathscr{N}(\mu_i, \sigma_i), \quad \sum_{i-1}^{N} w_i = 1, \qquad (6.69)$$

**Table 6.1** First 12 moments of mixture Gaussian distribution

| Moment | Value | Moment | Value |
|---|---|---|---|
| $\mu_1$ | 1.803323915192423 | $\mu_2$ | 5.618825039917679 |
| $\mu_3$ | 22.21194221994151 | $\mu_4$ | 98.56381003061070 |
| $\mu_5$ | 467.7931304312276 | $\mu_6$ | 2326.755938195312 |
| $\mu_7$ | 12010.60922300377 | $\mu_8$ | 63991.84370378382 |
| $\mu_9$ | 350667.3348484823 | $\mu_{10}$ | 1971324.861855061 |
| $\mu_{11}$ | 11345464.98539287 | $\mu_{12}$ | 66731974.28785729 |

where $\mathcal{N}(\mu, \sigma)$ is the normal distribution with mean $\mu$ and variance $\sigma$. The following mixture distribution with two terms is considered

$$\mathscr{P}(x) = \alpha(0.8\mathcal{N}(0.5, 1) + 0.2\mathcal{N}(4, 1))\mathbb{1}_{[0,8]}, \qquad (6.70)$$

where $\mathbb{1}_A$ is the box function which is equal to 1 if $x \in A$ and is zero otherwise. The parameters are chosen such that the final mixture is a bimodal distribution (see Behboodian [53] and Schilling et al. [54] for the necessary and sufficient conditions). The normalization constant, $a = 1.327743884718793$, ensures $\int_0^8 P(x)\mathrm{d}x = 1$. First 12 moments of this function are given in Table 6.1, which are calculated by direct integration using adaptive quadrature method [55] with relative and absolute tolerance of $1\mathrm{e}-8$ and $1\mathrm{e}-15$, respectively. Note that the PDF is normalized, and therefore, $\mu_0 = 1$.

Figure 6.3 shows the simulation of the mixture distribution using the Laguerre polynomials for 200 discrete points. More moments are needed in this example to correctly capture the tail of the distribution.

Table 6.2 shows the fractional moments of the mixture Gaussian distribution using Laguerre polynomials method (LPM) and DFMM. DFMM estimates the fractional moments particularly well as long as the number of integer moments is near the value of the fractional moments. For example, using first 3 integer moments, the estimated values for $\mu_{1.2}$, $\mu_{2.3}$, and $\mu_{3.4}$ are particularly precise with maximum error of 5.2 % for $\mu_{3.4}$, which are better than the estimates provided by LPM. However, for $\mu_{4.5}$, a large error is detected using the DFMM. Same calculations are performed using the first 5 integer moments, and the results are listed in Table 6.3. In this example, very precise estimates are provided using DFMM, which are all better than those calculated using LPM.

In the LPM case, increasing the number of moments to five actually increases the error. The test with seven and nine moments is performed, and a mild oscillatory convergence is detected, which is not reported in other studies using Laguerre series. On the other hand, DFMM is based on a firm mathematical ground with predictable behavior, which is a direct consequence of the short-memory principle discussed in Sect. 6.4.4. Parameter $\lambda$ can easily be adjusted to provide better results; for example, setting $\lambda = \frac{2}{p+1}$ results in $\mu_{4.5} = 209.9239$ with %Err $= 1.57$ (see Table 6.3).

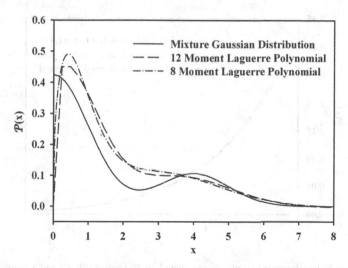

**Fig. 6.3** A mixture normal distribution—parameters in Eq. (6.70) are chosen such that the distribution is actually a bimodal distribution. The PDF is then reconstructed using 8 and 12 moments

**Table 6.2** Fractional moments estimated using three integer moments ($\mu_0$, $\mu_1$, and$\mu_2$)

| Moment | Value | LPM | DFMM | LPM %Err | DFMM %Err |
|---|---|---|---|---|---|
| $\mu_{1.2}$ | 2.1974 | 2.1093 | 2.1333 | 4.0093 | 2.9171 |
| $\mu_{2.3}$ | 8.3418 | 7.5806 | 8.5486 | 9.1249 | 2.4793 |
| $\mu_{3.4}$ | 39.9035 | 37.1028 | 37.8250 | 7.0187 | 5.2088 |
| $\mu_{4.5}$ | 213.2800 | 219.6070 | 140.6065 | 2.9665 | 34.0742 |

**Table 6.3** Fractional moments of a mixture Gaussian distribution, estimated using 5 integer moments ($\mu_0$, $\mu_1$, $\mu_2$, $\mu_3$, and$\mu_4$)

| Moment | Value | LPM | DFMM | LPM %Err | DFMM %Err |
|---|---|---|---|---|---|
| $\mu_{1.2}$ | 2.1974 | 2.2203 | 2.1804 | 1.0431 | 0.7741 |
| $\mu_{2.3}$ | 8.3418 | 8.7174 | 8.3543 | 4.5029 | 0.1501 |
| $\mu_{3.4}$ | 39.9035 | 45.4930 | 39.9454 | 14.0075 | 0.1050 |
| $\mu_{4.5}$ | 213.2800 | 271.8476 | 209.9239 | 27.4604 | 1.5736 |

### 6.5.3 Rice–Nakagami Distribution

A Rice–Nakagami distribution can be written by Majumdar and Gamo [56]

$$\mathcal{P}(x; I_c, \sigma) = \frac{1}{\sigma^2} \exp\left(-\frac{x + I_c}{\sigma^2}\right) I_0\left(2\frac{\sqrt{x I_c}}{\sigma^2}\right), \qquad (6.71)$$

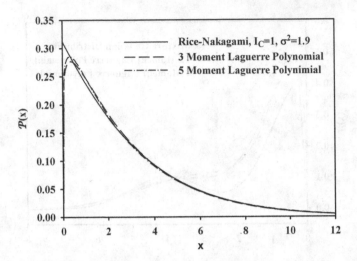

**Fig. 6.4** Rice–Nakagami distribution—reconstructed on 200 points using 3 and 5 moments

where $\sigma$ is the standard deviation and $I_c = \mu - \sigma^2$, with $\mu$ being the mean. $I_0$ is the zeroth-order modified Bessel function of the first kind. Higher-order moments are analytically given by Majumdar [57]:

$$\mu_j = \sigma^{2j} \exp\left(-\frac{I_c}{\sigma^2}\right) \Gamma(j+1) \mathscr{M}\left(j+1, 1, \frac{I_c}{\sigma^2}\right), \qquad (6.72)$$

where $\mathscr{M}$ is the confluent hyper-geometric function (Kummar function). Parameters $I_c$ and $\sigma^2$ are set to 1 and 1.9, respectively. Figure 6.4 shows the reconstructed PDF (Eq. 6.71) in [0, 12] for 200 discrete points with the normalization constant calculated to be 0.99994848708433 on this interval. Evidently, a Laguerre expansion can capture the features of this distribution with very high accuracy even with very limited number of moments. Note that only the first three and five moments are used to produce Fig. 6.4, whereas in Figs. 6.1 and 6.3, a larger set of moments was used. Figure 6.5 shows the exact values of the fractional moments and the values calculated using LPM and DFMM methods. Since this distribution can accurately be reconstructed using LPM, very accurate estimations for the fractional moments can be achieved with errors never exceeding 2.5 %. Despite this, DFMM still produces acceptable results with the maximum error of 22 % in calculating $\mu_{2.3}$ and error for $\mu_{1.2}$, $\mu_{3.4}$, and $\mu_{4.5}$ being in the same range as those calculated by LP method.

Estimation of the fractional moments using a constant value as suggested by Gzyl and Tagliani [50] is also attempted. The results are plotted in Fig. 6.5, and the only reliable results are those calculated between the available integer moments, i.e. $\mu_{1.2}$ and $\mu_{2.3}$. The values for $\mu_{3.4}$ and $\mu_{4.5}$ are severely underestimated and practically unusable. Note also that in this example, the maximum of the suggested interval is used, i.e. $H = 2$, using smaller values causes even larger underestimations. However,

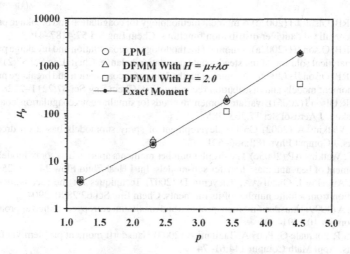

**Fig. 6.5** Fractional moments of the Rice–Nakagami distribution—the moments are calculated using LPM and DFMM methods using both an adaptive value and constant value for $H$

it should be stated that the adaptive value for $H$ is selected to work with very small number of integer moments, $3 \leq N \leq 5$, which has practical applications, for example, when solving Eulerian field equations. Therefore, if a larger number of integer moments are available in the model, one should use a more conservative value for $H$ or include the number of integer moments, $N$, in the definition of $\lambda$.

## 6.6 Summary

A basic introduction to general moment estimation is provided, and linear, non-linear and PDF reconstruction methods are considered. Relevant to the poly-disperse multiphase flow is the need to estimate non-integer and negative moments when only a few integer moments are available. MEMs are the method that is consistent with the underlying PDF framework under which the source terms are built. However, the method is slow and requires third-order moments to obtain non-trivial solutions. PDF reconstruction methods, presented here for $1D$ only, suggest a useful first step to obtain non-integer moments as a first guess for a MEM solver.

## References

1. Rigopoulos S (2010) Population balance modelling of polydispersed particles in reactive flows. Prog Energy Combust Sci 36:412–443
2. Frenklach M, Harris SJ (1987) Aerosol dynamics modeling using the method of moments. J Colloid Interface Sci 118:252–261

3. Diemer RB, Olson JH (2002c) A moment methodology for coagulation and breakage problems: part 3-generalized daughter distribution functions. Chem Eng Sci 57:4187–4198
4. Diemer RB, Olson JH (2002a) A moment methodology for coagulation and breakage problems: part 1-analytical solution of the steady-state population balance. Chem Eng Sci 57:2193–2209
5. Diemer RB, Olson JH (2002b) A moment methodology for coagulation and breakage problems: part 2 moment models and distribution reconstruction. Chem Eng Sci 57:2211–2228
6. Diemer R, Olson J (2006) Bivariate moment methods for simultaneous coagulation, coalescence and breakup. J Aerosol Sci 37:363–385
7. Beck J, Watkins A (2002) On the development of spray submodels based on droplet size moments. J Comput Phys 182:586–621
8. Beck JC, Watkins AP (2003c) The droplet number moments approach to spray modelling: the development of heat and mass transfer sub-models. Int J Heat Fluid Flow 24:242–259
9. John V, Angelov I, Oncul AA, Thevenin D (2007) Techniques for the reconstruction of a distribution from a finite number of its moments. Chem Eng Sci 62:2890–2904
10. Tagliani A (1999) Hausdorff moment problem and maximum entropy: a unified approach. Appl Math Comput 105:291–305
11. Inverardi P, Pontuale G, Petri A, Tagliani A (2003) Hausdorff moment problem via fractional moments. Appl Math Comput 144:61–74
12. Inverardi P, Pontuale G, Petri A, Tagliani A (2005) Stieltjes moment problem via fractional moments. Appl Math Comput 166:664–677
13. Pintarellia M, Vericat F (2003) Generalized Hausdorff inverse moment problem. Phys A 324:568–588
14. Talenti G (1987) Recovering a function from a finite number of moments. Inverse Prob 3: 501–517
15. Volpe EV, Baganoff D (2003) Maximum entropy PDFs and the moment problem under near-Gaussian conditions. Prob Eng Mech 18:17–29
16. Koopman BO (1969) Relaxed motion in irreversible molecular statistics. Stoch Process Chem Phys 15:37–63
17. Shannon CE (1948) A mathematical theory of communication. Bell Syst Tech J 27:379–623
18. Paris J, Vencovska A (1997) In defense of the maximum entropy inference process. Int J Approx Reasoning 17:77–103
19. Archambault MR, Edwards CF, McCormack RW (2003d) Computation of spray dynamics by moment transport equations I: theory and development. Atomization Sprays 13:63–87
20. Blinnikov S, Moessner R (1998) Expansions for nearly Gaussian distributions. Astron Astrophys Suppl Ser 130:193–205
21. Abramowitz M, Stegun I (1972) Handbook of mathematical functions with formulas, graphs, and mathematical tables. Dover Publications, New York
22. Majumdar AK, Luna C, Idell P (2007) Reconstruction of probability density function of intensity fluctuations relevant to free-space laser communications through atmospheric turbulence. In: Proceedings of SPIE
23. Kendall M, Stuart A, Ord J (1991) Kendall's advanced theory of statistics: distribution theory, vol 1. Wiley, New York
24. Gaztanaga E, Fosalba P, Elizalde E (2000) Gravitational evolution of the large-scale probability density. Astrophys J 539:522–531
25. Haeri S, Shrimpton J (2012) Closure of non-integer moments arising in multiphase flow phenomena. Chem Eng Sci 75(0):424–434. http://dx.doi.org/10.1016/j.ces.2012.03.052
26. Mood A, Graybill FA, Boes D (1974) Introduction to the theory of statistics. McGraw-Hill, New York
27. Scott SJ (2006) A PDF based method for modelling polysized particle laden turbulent flows without size class discretisation. Ph.D. thesis, Imperial College, London
28. Jaynes E (1957a) Information theory and statistical mechanics. Phys Rev 106:620–630
29. Sellens RW, Brzustowski TA (1985) A prediction of the drop size distribution in a spray from first principles. Atomisation Spray Technol 1:89–102

30. Sellens RW (1989) Prediction of the drop size and velocity distribution in a spray, based on the maximum entropy formalism. Part Part Syst Charact 6:17–27
31. Ahmadi M, Sellens RW (1993) A simplified maximum-entropy-based drop size distribution. Atomization Sprays 3:291–310
32. Boyaval S, Dumouchel C (2001) Investigation on the drop size distribution of sprays produced by a high-pressure swirl injector. Measurements and application of the maximum entropy formalism. Part Part Syst Charact 18:33–49
33. Dumouchel C, Boyaval S (1999) Use of the maximum entropy formalism to determine drop size distribution characteristics. Part Part Syst Charact 16:177–184
34. Archambault MR, Edwards CF, McCormack RW (2003a) Computation of spray dynamics by moment transport equations II: application to quasi-one dimensional spray. Atomization Sprays 13:89–115
35. Zwillinger D (2003) CRC standard mathematical tables and formulae. CRC Press, Boca Raton
36. Kreyszig E (1999) Advanced engineering mathematics. Wiley, New York
37. Alhassid Y, Agmon N, Levine RD (1978) An upper bound for the entropy and its applications to the maximal entropy problem. Chem Phys Lett 53:22–26
38. Alhassid Y, Agmon N, Levine RD (1979) An algorithm for finding the distribution of maximum entropy. J Comput Phys 30:250–258
39. Mustapha H, Dimitrakopoulos R (2010) Generalized Laguerre expansions of multivariate probability densities with moments. Comput Math Appl 60:2178–2189
40. Cody W (1976) An overview of software development for special functions. Lect Notes Math 506:38–48
41. Lebedev N (1972) Special functions and their applications. Dover Publications, New York
42. Ross B (1977) Fractional calculus: an historical apologia for the development of a calculus using differentiation and antidifferentiation of noninteger orders. Math Mag 50:115–122
43. Podlubny I (1999) Fractional differential equations. Academic Press, San Diego
44. Hilfer R (2000) Fractional calculus in physics. World Scientific, Singapore
45. Schiavone S, Lamb W (1990) A fractional power approach to fractional calculus. J Math Anal Appl 149:337–401
46. Cottone G, Di Paola M (2009) On the use of fractional calculus for the probabilistic characterization of random variables. Probab Eng Mech 24:321–334
47. Caputo M (1967) Linear models of dissipation whose q is almost frequency independent-II. Geophys J R Astr Soc 13:529–539
48. Cressie N, Borkent M (1986) The moment generating function has its moments. J Stat Plan Infer 13:337–344
49. Deng W (2007) Short memory principle and a predictor-corrector approach for fractional differential equations. J Comput Appl Math 206:174–188
50. Gzyl H, Tagliani A (2010) Hausdorff moment problem and fractional moments. Appl Math Comput 216;3319–3328
51. Alexiadisa A, Vanni M, Gardin P (2004) Extension of the method of moments for population balances involving fractional moments and application to a typical agglomeration problem. J Colloid Interface Sci 276:106–112
52. Lubich C (1986) Discretized fractional calculus. SIAM J Math Anal 17:704–719
53. Behboodian J (1970) On the modes of a mixture of two normal distributions on the modes of a mixture of two normal distributions. Technometrics 12:131–139
54. Schilling M, Watkins A, Watkins W (2002) Is human height bimodal? Am Stat 56:223–229
55. Shampine L (2008) Vectorized adaptive quadrature in Matlab. J Comput Appl Math 211: 131–140
56. Majumdar AK, Gamo H (1982) Statistical measurements of irradiance fluctuations of a multipass laser beam propagated through laboratory-simulated atmospheric turbulence. Appl Opt 21:2229–2235
57. Majumdar A (1984) Uniqueness of statistics derived from moments of irradiance fluctuations in atmospheric optical propagation. Opt Commun 50:1–7

# Appendix A: Averaging Operators

The averaging operators used in Jackson [1] are presented. For further details on the form of averaged time and space derivatives, the reader is referred to [1].

## A.1 Fluid Average

The weighting factor of Eq. 3.1 can be used to define the volume fraction of the fluid at position $\mathbf{x}$ according to

$$\alpha_f(\mathbf{x}) = \int_{V_f} g(|\mathbf{x} - \mathbf{y}|)d\mathbf{y}, \tag{A.1}$$

where the integration is over the entire fluid volume. The quantities are also time dependent, but this has been omitted for clarity. Similarly, the average of a general property $\psi$ (where $\psi$ is vector or scalar) is defined as

$$\alpha_f \langle \psi \rangle_f(\mathbf{x}) = \int_{V_f} \psi(\mathbf{y}) g(|\mathbf{x} - \mathbf{y}|)d\mathbf{y}. \tag{A.2}$$

The average of space and time derivatives is slightly more involved, and the reader is once again referred to [125] for details.

## A.2 Solid Average

The solid averages are analogous to the fluid-phase averages and are constructed by integration over the entire volume occupied by the solid particulate phase. Thus, the volume fraction of the solid is given by

J. S. Shrimpton et al., *Statistical Treatment of Turbulent Polydisperse Particle Systems*, Green Energy and Technology, DOI: 10.1007/978-1-4471-6344-2, © Springer-Verlag London 2014

$$\alpha_s(\mathbf{x}) = \sum_n \int_{V_s} g(|\mathbf{x} - \mathbf{y}|) d\mathbf{y}, \tag{A.3}$$

and similarly, the average of a solid-phase property is defined as

$$\alpha_s \langle \psi \rangle_s (\mathbf{x}) = \sum_n \int_{V_s} \psi(\mathbf{y}) g(|\mathbf{x} - \mathbf{y}|) d\mathbf{y}. \tag{A.4}$$

## A.3 Mixture Average

A mixture or overall volume average $\langle \psi \rangle$ of a point property $\psi$ is defined as

$$\langle \psi \rangle_m (\mathbf{x}) = \int_V \psi(\mathbf{y}) g(|\mathbf{x} - \mathbf{y}|) d\mathbf{y}$$

$$= \alpha_f(\mathbf{x}) \langle \psi \rangle_f + \alpha_s(\mathbf{x}) \langle \psi \rangle_s, \tag{A.5}$$

or alternatively, the mass weighted average is given as

$$\rho_m(\mathbf{x}) = \rho_f \alpha_f \langle \psi \rangle_f + \rho_s \alpha_s \langle \psi \rangle_s, \tag{A.6}$$

where $\rho_f$ and $\rho_p$ are the material densities of the fluid and particulate phases, respectively, and $\rho_m(\mathbf{x})$ is the mixture density.

## A.4 Particle Average

The particle phase is a discrete phase, so the averaged quantities require a different averaging method. The particles to be averaged can be completely described according to the 'point-particle' assumption [99]. Following this assumption, the motion of each particle is determined by the particle velocity and resultant force (Jackson also considers the moment acting on the particle). The details of the stress distribution within the particle are not required. The number of particles per unit volume $n$ at position $\mathbf{x}$ is defined as

$$n(\mathbf{x}) = \sum_{i=1}^{n_p} g(|\mathbf{x} - \mathbf{x}^{(i)}|), \tag{A.7}$$

where $\mathbf{x}^{(i)}$ is the position of the centre of an individual particle $i$. Similarly, the average of a property $\psi$ is given by

$$n(\mathbf{x})\langle\psi\rangle_{\mathbf{p}} = \sum_{i=1}^{n_p} \psi^{(i)} g(|\mathbf{x} - \mathbf{x}^{(i)}|). \qquad (A.8)$$

# Reference

1. Jackson R (1997) Locally averaged equations of motion for a mixture of identical spherical particles and a Newtonian fluid. Chem Eng Sci 52:2457–2469

# Index

Printed in the United States
By Bookmasters